TURNABOUT AND DECEPTION

TURNABOUT AND DECEPTION
Crafting the Double-Cross and the Theory of Outs

BARTON WHALEY

Edited by Susan Stratton Aykroyd

Naval Institute Press
Annapolis, Maryland

Naval Institute Press
291 Wood Road
Annapolis, MD 21402

© 2016 by U.S. Naval Institute
All rights reserved. No part of this book may be reproduced or utilized in any form or by any means, electronic or mechanical, including photocopying and recording, or by any information storage and retrieval system, without permission in writing from the publisher.

Library of Congress Cataloging-in-Publication Data
Names: Whaley, Barton, author. | Aykroyd, Susan Stratton, editor.
Title: Turnabout and deception : crafting the double-cross and the theory of outs / Barton Whaley ; edited by Susan Stratton Aykroyd.
Other titles: Crafting the double-cross and the theory of outs
Description: Annapolis, Maryland : Naval Institute Press, [2016]
Identifiers: LCCN 2016015946| ISBN 9781682470282 (alk. paper) | ISBN 9781682470299 (mobi)
Subjects: LCSH: Deception (Military science)—Case studies. | Military intelligence.
Classification: LCC U167.5.D37 W38 2016 | DDC 355.4/1—dc23 LC record available at https://lccn.loc.gov/2016015946

♾ Print editions meet the requirements of ANSI/NISO z39.48–1992 (Permanence of Paper).
Printed in the United States of America.

24 23 22 21 20 19 18 17 16   9 8 7 6 5 4 3 2 1
First printing

Editor's note: This book combines two monographs written by Dr. Barton Whaley (1928–2013) for the Foreign Denial & Deception Committee, originally titled *Turnabout: Crafting the Double-Cross* and *When Deception Fails: The Theory of Outs*, respectively.

Table of Contents

Introduction by John McLaughlin. vii

Volume I. *Turnabout* . . . . .I-i
Contents . . . . I-vii
Executive Summary. . . . . .I-ix
1. Introduction. I-1
2. Defining Turnabout I-3
3. The Three Steps to Turnabout. I-11
4. Types of Turnabout I-13
5. The Security of Options I-73
6. Double or Not?: Two Parallel Cases I-80
7. Double-Agent Systems: Dead or Alive?. I-102
8. Limits of the Game of Turnabout: Simple Cycles or Infinite Layers?. . .I-103
Appendix: The Double-Crosser's Dictionary I-106
Bibliography. I-113

Volume II. *When Deception Fails* . . . . II-i
Contents . . . . .II-vii
Introduction. . . . . II-ix
Summary of Findings & Conclusion . . . . II-xi
Part One: Problems of Failure II-1
1. Causes of Failure II-1
2. Degrees & Types of Failure II-24
3. Coping in Military Practice II-73
Part Two: Solutions to Failure II-77
4. Coping in Theory. II-77
5. Military Appreciation of Outs II-81
6. Priorities for Outs II-83
7. Beyond the Out: Asymmetries between Deceiver & Target II-86
Bibliography. II-91
Appendix: Extracts from 1982 CIA contract paper II-98
Indices. II-105

Introduction

The author of these volumes—*Turnabout* and *When Deception Fails*—has always been one of my heroes, for more reasons than readers may immediately know.

We share three passions. First, we both had a long association with the intelligence profession. In Dr. Whaley's case this was as an expert on the art of deception and as an advisor to intelligence officers who are its most likely targets and, potentially, its victims. In my case, this was as a CIA officer for more than three decades, working on nearly every aspect of the business, and serving during my last four years as the Agency's deputy director and briefly as its acting director.

Second, and this is the lesser known association, Dr. Whaley and I shared a passion for what in various forms is called conjuring, legerdemain, or prestidigitation—or, as more commonly known, magic. In my case I caught the bug at the tender age of eleven and have never been able to shake it. We shared membership in the world's largest magic fraternity, the International Brotherhood of Magicians. Dr. Whaley wrote a number of landmark books on the subject, and I have been, as time would permit in the midst of a busy intelligence career, a performer and lecturer on magic.

Third, Dr. Whaley and I both saw close connections between the art of intelligence deception and the sort of deception practiced by professional magicians. There is evidence of Dr. Whaley's conviction on that score in these volumes, as when he refers to the role that British magicians, in particular, played in some of the great deceptions directed at Hitler's forces in World War II. Dr. Whaley in much of his work mentions also the brilliant deception work of British colonel Dudley Clarke, who headed the legendary A-Force, the primary deception workshop during that conflict. In my own research, I discovered that Clarke was the grandson of Sidney W. Clarke, an early official of Britain's Magic Circle (the first major magic society in the West), and the author of *Annals of Conjuring*—to this day the most authoritative history of early magic in the English language. Young Clarke is said to have received lessons in conjuring from his grandfather, and this cannot have but influenced his later thinking about deception on a grander scale.

I cannot lay claim to anything approaching the written scholarship on magic that Dr. Whaley produced. But I have lectured on several occasions to magic societies on what I call the "kindred arts of magic and espionage" and the role of magic at key moments in history—putting the art of deception at the service of the state.

Every phase of the intelligence discipline involves some activity that finds an analogue in the arts of magic and deception. Magic so often involves more psychology than digital dexterity—planting suggestions and creating mindsets that lead the audience to draw unwarranted conclusions about how something was accomplished. Similarly, intelligence analysts are often defeated by mindsets that fit data into unwarranted patterns leading to the wrong conclusions. Simply being aware of how magic succeeds can sensitize an analyst to the danger that lurks in often misleading data sets.

For the operations officer, the challenge is different but also finds an echo in the magic field. An operations officer in a hostile environment marked by heavy and continuous surveillance faces the challenge of carrying out his or her work without being discovered even while being constantly watched. Once again this is akin to the challenge the magician faces on a brightly lit stage seeking to entertain with deception an audience determined to detect his or her tricks. The magician must do things in front of an audience to accomplish his or her "mission" without the audience noticing what he or she is doing. This is not all that different from an intelligence officer on a street under surveillance planning to pass a message or place a package while being watched. The case officer who can think like a magician starts off with an advantage, and any case officer reading Barton Whaley's books will get a sophisticated introduction to that kind of thinking.

Lest the reader think Whaley is dealing with abstract theories, it is important to note that in both volumes he explicitly seeks to relate his analysis and recommendations to the most pressing and practical contemporary challenges. Nothing has occupied the resources and creativity of American and allied military and intelligence services over recent decades more than the challenge of dealing with insurgency and terrorism. Whaley goes right to the heart of these problems by pointing out that insurgents and terrorists win primarily through asymmetric techniques. That is, practices that effectively checkmate the vast advantages that large conventional armies carry.

To be sure, he notes, nations facing these challenges have developed counterinsurgency (COIN) and counterterrorism (CT) doctrines to deal with asymmetric challenges. But as an expert on deception, he properly notes that once such doctrines are in place, they "too often become counterproductive straightjackets" that limit the flexibility his deception techniques call for. Smart deception techniques can essentially match asymmetry with asymmetry—especially the sort of the techniques Whaley defines in the *Turnabout* book.

Typically the idea implied by the term turnabout is the classic "double agent"—something intelligence specialists have written about since Chinese strategist Sun Tzu's classic text *The Art of War* in the sixth century BC.

Whaley's breakthrough is to give concreteness to Sun Tzu's idea by laying out specific models for different types of "double crosses." These range from what he calls Hiding Under Thin Cover to Hiding in Plain Sight to my personal favorite: Double-Bluffing. The latter's appeal for me lies in its near perfect analogy to a conjuring art form known by the faintly insulting term "sucker trick." This is the kind of magical deception in which the performer leads audience members subtly (or sometimes boldly) to believe that they have discovered the explanation for some miracle, only to reveal in the trick's denouement that they were completely wrong. In well-constructed stage magic, this will leave the audience laughing and astounded. In military and intelligence operations, the result is devastating.

Whaley's love for magic infuses both these books but is most explicit in *When Deception Fails*. While this book seeks to diagnose the potential causes of failure (it lays out five) and offers strategies for avoiding it, its key message lies in another idea: failure is always a possibility and the only guarantee against it is to always have a contingency plan—or what a magician (and Whaley) calls an "out." That means literally what it implies—some stratagem that ensures you can bring the deception to an acceptable conclusion—a way "out" of failure.

Many magic tricks are high wire acts, at least figuratively. Myriad things can go wrong. You can drop the cards or simply get mixed up in the course of some complicated manipulation or procedure—things that are designed to look simple to an audience but actually have many "moving parts," so to speak. So prone are magicians to this type of mishap that there is even a book explicitly written to help magicians prepare for and avoid failure—*Outs, Precautions, and Challenges* by Charles H. Hopkins (cited by Whaley in his extensive bibliography).

In the volume you hold, Whaley has written the intelligence and military counterpart to Hopkins' book. Deception operations in the military or intelligence worlds are also high wire acts with many moving parts. Whaley lays out how the theory of "outs" can be applied to military deceptions in particular by simply making sure that any deception operation is designed so that it can shift direction if discovered. This may be something as simple as holding some forces in reserve that can be used at the last minute to shift the focus of an attack. Whaley provides an actual formula involving nine attributes of a deception operation that a commander can manipulate—each of which can have built-in alternatives. These range from timing to style, and ultimate intention. In this he follows the advice of Britain's master deception planner Dudley Clarke, who taught his units to "always leave . . . an escape route."

As the twenty-first century unfolds, it is clear that the United States will have to navigate in a vastly more complex and competitive world. Barton Whaley with these volumes places in the hands of national policymakers tools that could prove of enormous value to national decision makers seeking to protect and advance American interests. They are truly gifts to the military and intelligence professions—at precisely the moment when they are most needed.

John McLaughlin was Deputy Director and Acting Director of Central Intelligence from 2000–2004 and now teaches at the Johns Hopkins School of Advanced International Studies (SAIS).

Volume I

TURNABOUT:
Crafting the Double-Cross

BARTON WHALEY

Editor: Susan Stratton Aykroyd

Foreign Denial & Deception Committee
National Intelligence Council
Office of the Director of National Intelligence
Washington, DC

January 2010

The views herein are the author's and not necessarily those of the Foreign Denial & Deception Committee

Volume I

TURNABOUT:
Crafting the Double-Cross

BARTON WHALEY

> That is good deceit which mates him first that first intends deceit.
> — Shakespeare, *Henry VI* (c. 1591), Part II, Act III, Scene I

> It is a double pleasure to deceive the deceiver.
> — Jean de la Fontaine, *Fables*, Book II, Fable 15, "The Cock and the Fox" (c. 1694)

> Turn about was fair Play.
> — Captain Dudley Bradstreet, *The Life and Uncommon Adventures* (1755), 338

In warm memory of

Amrom Harry Katz

(1915-1997)

Modest physicist,
collector of counter intuitive toys,
world-class practical joker,
champion of Dr. R. V. Jones,
wise deception analyst,
pioneer IMINT intelligencer,
advocate of simple solutions to high-tech problems,
and provocateur of lazy minds.

Contents

Executive Summary... I-ix
    Main Finding.. I-ix
    Five Insights .... I-x
    Three Tools .... I-xiii
    Five Recommendations....................................... I-xiii
    Main Conclusion.. I-xvi

1. Introduction... I-1
2. Defining Turnabout... I-3
    2.1. Turnabout = Double-Cross I-5
    2.2. Cross & Double-Cross I-6
    2.3. Deception & Counterdeception I-6
    2.4. Parthian Shots & Parting Shots I-7
    2.5. Flag & False Flag I-8
    2.6. Blowback vs Playback.................................... I-9
    2.7. Bluff & Double-Bluff..................................... I-9
    2.8. Time Distortion Effects I-10

3. The Three Steps to Turnabout................................. I-11
    3.1. Deception I-11
    3.2. Detection I-12
    3.3. Turnabout I-12

4. Types of Turnabout I-13
    4.1. Double-Ambush & Counter-Ambush.................. I-14
    Case 4.1.1: The Warrenpoint Double-Ambush, Northern Ireland, 1979..... I-15
    Case 4.1.2: A Magician Counter-Ambushes the Afghans, 1902............ I-17
    Case 4.1.3: Counter-Sniping, 1915-2009................................ I-19
    4.2. Double Back... I-20
    Case 4.2.1: The Keyhole Satellite that Played Dead, 1977-78............. I-20
    Case 4.2.2: The D-Day MOONSHINE Electronic Double Back, 1944 ..... I-21
    Case 4.2.3: The Most Dangerous Game — in Fiction, 1924................ I-22
    4.3. Double Exchange I-24
    Case 4.3.1: The Chinese 7th Stratagem: Make the Illusion Real, AD 755..... I-25
    Case 4.3.2: The Alamein MUNASSIB Switch, 1942...................... I-26
    Case 4.3.3: The Actor Tricks the Magicians, 1922........................ I-28
    4.4. Hiding in Plain Sight I-29
    Case 4.4.1: Indomitable Jones and the Malta Radar I-30
    Case 4.4.2: Dudley Clarke and the Aircraft Dummies I-31

- 4.5. Hiding under Thin Cover I-32
  - Case 4.5.1: The FAREWELL Gambit, 1982-84 .... I-32
  - Case 4.5.2: HEINRICH Helps Set the Stage for the Battle of the Bulge, 1944 . I-34
  - Case 4.5.3: The Bar Code Orange Con, 2003-2009........................ I-35
- 4.6. Provoking Truth I-47
  - Case 4.6.1: Plan GISKES: Mr. Marks Unmasks Col. Giskes, Feb 1943 .... I-47
  - Case 4.6.2: The Dreyfus Ruse, 1899 .... I-49
  - Case 4.6.3: The Forensic Accountant vs the Embezzler, 1970s I-50
  - Case 4.6.4: America's Midway Ruse, 1942 .... I-50
  - Case 4.6.5: Hypothetical Scenarios, 1984 to present..................... I-52
- 4.7. Double-Bluffing I-53
  - Case 4.7.1: The Booby Trapped Booby Trap: Major Foot's Luck, 1943........ I-53
  - Case 4.7.2: The CORONA Satellite That Wasn't, 1958 .... I-54
  - Case 4.7.3: Jones and the Fake Jay Beam, 1941.......................... I-55
  - Case 4.7.4: Jones and the Double-Telltale Envelope, early 1940s I-56
  - Case 4.7.5: The Allies Double-Bluff the German Gothic Line in Italy, 1944.... I-57
  - Case 4.7.6: RFE Turnabouts a Czech Terror Plot, 1957..................... I-59
- 4.8. Double Agentry I-60
  - Case 4.8.1: CHEESE: World's First Double-Agent System, 1939-1944......... I-61
  - Case 4.8.2: NORTH POLE: German Playback Double-Agentry,1942-1944 .. I-62
  - Case 4.8.3: LAGARTO: Japanese Double-Agentry, 1943-45 .... I-64
  - Case 4.8.4: SCHERHORN: A Soviet Playback Operation, 1944-1945......... I-64
  - Case 4.8.5: X-2: The American Way of Double-Agentry I-65
- 4.9. Tripwires & Telltales.. I-65
  - Case 4.9.1: Daniel & the Priests of Bel, c.100 BC I-66
  - Case 4.9.2: Noor's Secret Triple Security Checks, 1943...................... I-67

5. The Security of Options I-73
   - Case 5.1: The No M.O. I-75
   - Case 5.2: The Secret M.O. I-76
   - Case 5.3: Sherman's March to Atlanta, 1864 .... I-76
   - Case 5.4: The Two Patton's Ruse, 1944 .... I-77

6. Double or Not?: Two Parallel Cases I-80
   - Case 6.1: GARBO, 1944 .... I-81
   - Case 6.2: Marwan, 1973-2007 .... I-84

7. Double-Agent Systems: Dead or Alive? I-102

8. Limits of the Game of Turnabout: Simple Cycles or Infinite Layers? . I-103

Appendix: The Double-Crosser's Dictionary I-106

Bibliography I-113

Executive Summary

This paper is the first systematic and cross-disciplinary analysis of how we can turn an opponent's attempted deception of us back upon them.[1] That's the rare — surprisingly rare — craft of turnabout or the double-cross. The focus here is on its practice in military and intelligence contexts. However, the 38 case-study examples in this paper also draw freely from other fields & disciplines in order to show that deception & counterdeception comprise a general psychological battle of wits. Moreover, this inter-disciplinary approach is a fruitful stimulus to identifying some general principles and primitive theory that can be applied to military and political-strategic situations.

Main Finding

While the notion of turnabout or deceiving the deceiver is often mentioned in the literature, this is the first study to develop a model of the different types or categories of turnabout or double-crossing. The analysis identified 9 such categories:

- COUNTER-AMBUSH:
  Set an ambush for one's attempted ambusher.

- DOUBLE-BACK:
  Return geographically to an earlier place or operationally to a discarded method.

- DOUBLE EXCHANGE:
  Have the dummy switch places with the real or vice versa.

- HIDING IN PLAIN SIGHT:
  Operate in such a "natural" way as to remain unnoticed.

- HIDING UNDER THIN COVER:
  Operate with minimal disguise or camouflage.

- PROVOKING TRUTH:
  Evoke accurate information by tricking the liar.

- DOUBLE-BLUFFING:
  Let the opponent see clues to a dummy ("notional") deception to distract them from the real method.

---

1   The only previous references to theory or principles of military deception turnabout, although sketchy & unsystematic, first appeared in the work of Harris (1973), Part III; and Whaley (1969), Chapter 4.3, which had been inspired by discussions with Harris in 1968-69.

- **DOUBLE AGENTRY:**
  Secretly coopt ("turn") an opponent's secret agents and then work them back against their original employer.
- **TRIPWIRES & TELLTALES:**
  But only when they are used as "silent alarms" that a) don't alert the stealthy intruder; and b) permit the forewarned occupier to take some countermeasure to trap, ambush, or otherwise surprise the surpriser.

Other categories seem likely. Plausibly, for example, Hiding Under *Deep Cover*. Further research & analysis should reveal them and/or refine the existing categories. The practical application of a comprehensive set of categories such as this is a welcome checklist for our deception analysts, planners, teachers, and theorists.

Five Insights

Insight 1
Turnabout or double-crossing operates at one level higher, one step above simple deception. Even so, like simple first-level deception, it's still a much less "tangled web" process than generally assumed.

An elegant & simple circular model is proposed that seems to give the double-cross system a more rigorous description & systematic explanation than the popular infinite-regression or "wilderness of mirrors" model. This finding represents a generalization to all types of turnabout of the researcher's earlier model that applied only to the single category of double agentry. (Chapter 8)

Insight 2
Turnabout or double-cross is by definition a form of asymmetric conflict. It's a "game changer" in every battle of wits. But not just in the usual sense of having introduced a more powerful weapon, a more effective tactic, a more closely coordinated command-and-control network, or some higher-tech surveillance device. Instead, turnabout introduces a whole new set of rules — your rules.

The very term "conventional warfare" says it all. It literally means making war according to convention — to a set of *mutually agreed* rules. Consequently, — the fog of war notwithstanding — outcomes can at least be assigned probabilities. In other words, symmetrical battles are more-or-less predictable. And, whenever the weak confront the strong, the outcome is highly predictable.

Terrorists, insurgents, and guerrillas have learned this the hard way. To even survive, much less win, they must engage their enemy with asymmetric strategies and tactics. In practical terms, this means changing the rules of engagement. Mao Zedong understood this when he said, "You fight your way and I'll fight my way." He did and he won — won all of China. And it is particularly salient, even ironic, that he formulated this aphorism during a March 1965 interview with a visiting delegation from a terrorist organization, the Palestine Liberation Organization.

Con artists and magicians are the only professionals who thrive on the we-make-our-own-rules principle. Only they take it as their prime directive and deploy it at least to some extent in every operation.

All other professionals use deception to greater or lesser degree, but the only ones that ever resort to double-crossing are the occasional Machiavellian politician, cut-throat businessmen, counter-espionage officers, and the rare stratagemic military commander. All of these fight by indirection, by imposing their own asymmetric strategy & tactics on their opponents.

Turnabout — as opposed to simple ground-level deception — is even rare in games, which gave us the word "turnabout"; fairly rare in sports, which gave us the words "cross" & "double-cross"; and almost non-existent on the battlefield.

One perceptive military commentator recently concluded that, "What the United States fears most is an adversary who plays by his own rules."[2] I don't agree. We don't so much fear them as that we are confused and baffled. They've chosen to fight us from the shadows, lying in covert ambush with IEDs or striking out of nowhere with commando-type raids or suicide bombers. Conventional thinking like "Why don't they come out and fight" and conventional responses by escalating conventional military force guarantee frustration and risk costly defeat. Since WW2 we've seen the evolution of some effective but partial solutions to asymmetric combat in various counter-guerrilla, counter-insurgent (COIN), and counter-terrorist methods. Fine. But these techniques and tactics have come down to us as rather narrowly defined doctrines. And doctrine too often becomes a counterproductive straitjacket, despite pious talk of "flexibility", "feeling our way", or "learning by trial and error".

Escape from this dilemma and a way forward is by recognizing the dominant principle involved. For this we must step back from the intricate details of asymmetric combat far enough to see its basic and simple architecture.

---

2   Lu Gang, "Tactical Openness: Open Measures Conceal Covert Tricks" *Zhongguo Guofang Bao* [thrice weekly newspaper sponsored by the PLA Daily], 8 Jun 2004.

The essence of asymmetric conflict, as with double-cross, is to let the opposition play its own game while we play ours. And, ideally, without the opponent realizing what's happening. Simple deception involves only twisting the rules or by taking advantage of secret information. Even the most elaborate deceptions only deflect the conventional rules, turning them to one side.

Conversely, because turnabout or double-cross is literally a contrarian art, it completely rewrites the rules, turning them 180-degrees, to become an entirely new game.

What if turnabout fails? Not just plain cheating or simple deception but flagrant double-dealing betrayals or playbacks. Obviously, failure can happen, although the cases studied and much anecdote suggest that abject failure is rare (see CASE 4.7.6). When failure does happen, the only common downsides are in the practitioner's acquiring either a reputation for double-dealing or a new disgruntled enemy or two. This reputation for "not playing by the rules" is, of course, one that magicians wear with pride and con artists flee by moving on to find an ever new bunch of suckers. The only downside for military commanders and counter-espionage officers is that the double-cross may become recognized as their M.O. (CASE 5.1).

But when payback succeeds, the payoff for the weaker combatant is a greatly enhanced chance of success; for the stronger it promises success at a much lower cost in human and material treasure. Is this just idle theorizing? No, as demonstrated by all of this paper's 38 case studies.

Insight 3

Part of "knowing your enemy" is knowing their M.O. If their M.O. includes any ability for turnabout play, strong hints of it may be available to the deception analyst. If they are magicians, con artists, or double agents, their M.O. will almost certainly include turnabout (double-crossing or a sucker gag), in which case it will be an obvious yellow or even a blatantly red flag. But only when the analyst has been trained to recognize those yellow & red flags. (Section 4.5.3; and Chapter 6).

Insight 4

A plausible, perhaps compelling, argument that, contrary to general belief, the revivals of WW2-type double-agent systems are quite likely projects. On smaller scales perhaps, but well worth the planners or analysts consideration. (Chapter 7)

Insight 5

Because our current Western military doctrines & organizations still generally undervalue & seldom use sophisticated deception, they artificially limit the rich possibilities for counter-deception of the turnabout type.

Three Tools

T1

An outline of a general procedure for the PROVOKING TRUTH category (Chapter 4.6).

T2

A preliminary — very preliminary — analytical technique for assessing the likelihood that someone might be a double agent (Chapter 6).

This is a simple rule-of-thumb (heuristic) that takes advantage of the fact that double agentry is a sub-category of con artistry. Consequently, many of the "red flags" that warn us of the possible presence of a confidence game operator (Section 6.2) also apply to a double agent (Section 6.1).

T3

Timeline Deception Analysis (TDA) is introduced as a useful tool for detecting deception, particularly when double-crossing is suspected. Details below in R4.

Five Recommendations

Consequently and finally, this paper generated five specific recommendations:

R1

Replicate & enlarge the data base of case studies involving turnabout.

This topic is sufficiently important to justify further research — replication & expansion by other analysts or even a team of analysts. Such an enhanced study would greatly benefit from a systematic search for additional relevant case studies. This could be a useful teaching aid.

R2

Attempt to expand the 8-category model of double-cross developed in this paper.

It would substantially enhance research and teaching to develop a more systematic set of double-cross or turnabout categories than the eight given here. These eight had evolved on a rough anecdote-by-anecdote basis only. That is, at best, a stopgap approach, a start-up that needs a rigorous follow-through.

R3

Develop a biographical data base of opponents' personnel engaged in political-military deception —one focused on their deception styles or MO.

This paper identified a specific need for the systematic collection & analysis of biographical intelligence on personnel who are specifically engaged in political-military deception at the planning, operational, and analysis stages.

This project should be of only slightly lower priority than the usual biographical profiling & gaming of opposing national leaders and military commanders. Those studies attempt to identify the opponent's weak points ("vulnerabilities") to be exploited or strengths to be avoided. In the proposed research, the focus would be specifically to identify the opponent's preferred deception M.O. If successful, our deception analysts will sharply, efficiently, & profitably reduce the range and number of "competing hypotheses" they need to consider & weigh.

See Chapter 8 on the potential for a general battle of deception wits. And see CASES 4.3.3, 5.1, and 5.2 for the specific "No M.O." and the "Secret M.O." games that can be played by deceivers.

R4

Further develop & apply Timeline Deception Analysis (TDA) as a highly effective tool for detecting deception, particularly when double-crossing is suspected.

Timelines are a potent way of showing who knew what when. This approach is clearly the most effective technique for overcoming the 20/20 hindsight problem. In other words, chronology introduces & enforces a kind of simulated naivete, freeing us from the straightjacket of hindsight bias to permit a more-or-less fresh & unprejudged intelligence analysis.

Then, if we combine chronology with our search for those yellow and red flags that openly signal deception, we obtain a potent formula for successful deception analysis. Timeline analysis is, of course, no more than the historian's basic procedure of organizing data into chronological sequence. But red-flagging is best identified through the theory & procedures of Incongruity

Analysis that originated with R. V. Jones, made explicit by William R. Harris, developed by Whaley & J. Bowyer Bell, and explicitly perfected by Frank Stech & Elsaesser.[3]

More importantly, chronological analysis can be almost as effective for discovering & unmasking deception in ongoing or on-line situations as it is for historical case studies. Even when limited to open sources, rough versions of red-flag analysis have successfully unmasked ongoing political-military deceptions.[4] In other words, Timeline Deception Analysis can be useful in detecting current deceptions and thwarting an imminent surprise.

The power of chronological analysis for deception analysis was first demonstrated in the author's doctoral thesis where it revealed the previously unappreciated relevance of deception in Operation BARBAROSSA, the German surprise invasion of Russia in 1941. That research method then substantially assisted in identifying dozens of deception cases in my *Stratagem: Deception and Surprise in War* (1969). However, it's only been in the course of researching & writing this present paper that I fully understood the originality of my particular Timeline Deception Analysis (TDA) procedure — specifically in CASE 4.5.3 and the 2 cases in Chapter 6 which focused on timelines plus red flagging.

R5

Develop a Deception Layer Model.

An interesting question lurks in the first of the 4 "Insights" given above and in Chapter 8. That insight is that double-crossing operates one level or layer higher than simple deception. The question is, what other levels of deception are out there? How many levels? Which types of deception are in each? How do we identify & define them? And can we then use them to rate the competency of both deceivers and deception analysts? Is it true, as I suspect, that in the battle of wits, two rules will hold:

- **Hypothesis 1:** The deceiver who operates at least one level higher than the opponent will usually succeed in deceiving.

- **Hypothesis 2:** The deception analyst who operates at least one level higher than the opponent will usually detect the deception.

If true, considerable predictive power could flow from a Deception Layer Model. I strongly recommend examining this question.

---

3   See Whaley, *Textbook of Political -Military Counterdeception* (FDDC, Aug 2007), Chapter 4 ("How to Detect"), 43-58.
4   As first systematically demonstrated by Ladislav Bittman, *The KGB and Soviet Disinformation* (Washington, DC: Pergamon-Brassey's, 1985)

Main Conclusion

Turnabout is a grossly underused tactic for countering our various opponents' deceptions.

This was the one clear and overarching conclusion produced during the research for this paper. While turnabout has been rarely used in the past, it needn't be in the future. The case examples collectively suggest that their historical rarity results mainly from a widespread failure of imagination. That can be corrected by providing our intelligence and deception analysts with a toolkit comprised of a checklist of turnabout ploys suited to countering specific types of deception operations that might be deployed against us.

Obviously, all the turnabout tricks described here are also known to our allies, opponents, and potential enemies. And some have proven themselves more proficient than we in crafting the double-cross. To win this battle of wits, our analysts and planners need to be properly armed. This paper has forged some of those weapons. Those need to be better honed. And more are needed.

Remember that turnabout is not the simple game where an aggressor initiates a deception against another deceiver. Instead, turnabout is a hyper-game where a defender interrupts the aggressor's deception and turns it back upon the other.

1. Introduction

> I assure you ... fallere fallentem non est fraus; or,
> in English, 'Tis no sin to cheat the Devil.
>
> — Daniel Defoe, "Tom a Bedlam's View of Bribery" (1722)

This paper attacks & answers the question, "What can be done to turn an opponent's deceptions back upon them?" When in early 2009 this highly specialized topic was first proposed as an FDDC study, one veteran DIA analyst labeled it, "A topic screaming for coverage and for suggestive and creative ways of applying this art to contemporary war/intelligence dimensions."

I agreed, having been sifting data on this question since 1968. My initial focus had been on military and political-strategic deception. After analysis of over a hundred cases, I grew puzzled by the relative passivity of the victims. I understood their surprise, even their anger. But why had so few victims taken any positive measures to a) detect the presence & nature of deception-based surprises beforehand, much less b) limit the subsequent damage of those surprises?

And why, when attacked, was their response too often only to counterattack — a rage-driven mindless thrashing back? These victims' mantra was simply fall back, regroup, bring up the reserves, and hit back. That's what every standard Western field manual told them. And the result? Seldom victory. Sometimes stalemate. Usually costly. And always catastrophic whenever a clever opponent had set a second trap — the old Parthian Shot trick of feigning retreat to sucker your counterattack into an ambush.

The literature on political, military, and espionage deception is extensive. But, although rich in anecdotes and case studies, it's surprisingly thin on theory or even basic principles — indeed very thin. Moreover, because we are forced to use only those cases where official or unofficial declassification has disclosed both sides of the deception equation, the cases in this paper are neither as many nor as recent as would be ideal. So, of necessity, to supplement these gaps in both theory & recency, I sought wider insight from among the practitioners of other disciplines. These proved to be mainly magicians with some additional material from con artists, professional gambling cheats, cognitive psychologists, and the few well-documented military examples and the more numerous counterespionage cases.

A brief explanation in owed any readers who may find this paper overloaded with its author's personal references and apparent name dropping. This is not some textbook that collects, summarizes, and rewrites the scattered works of others. Instead, it is largely a combination of original research and past collaborations with and inspirations from a few fully credited colleagues. Consequently, I accept the famous recommendation of Nobel Prize physiologist Sir Peter Medawar.[5] He urged full disclosure of the first-person history of each research project rather than mindlessly following the shallow tradition of all scholarly publications that, by depersonalizing, tends to conceal biases and hide the pitfalls & blind alleys in the research. The conventional format inevitably conveys a misleading appearance of tidy academic infallibility. In most instances this is merely a bad habit of benign oversight. Unfortunately, particularly in the less rigorous sciences, it is a format that is often unwittingly manipulated to conceal incompetence, charlatanism, and outright fraud.

This paper not merely addresses & describes the craft of turnabout but answers or half answers several puzzling questions about it. That it also identifies & limelights new questions may even be its best feature.

---

5 P. B. Medawar, "Is the Scientific Paper a Fraud?," *The Listener*, Vol.70 (12 Sep 1963), 377.

2. Defining Turnabout

> "Whosoever thou art that dost another wrong, do but turn the tables: imagine thy neighbour were now playing thy game, and thou his ...."
>
> — Robert Sanderson, *Twelve Sermons* (London: 1634), Sermon VI

Neither the English language — nor any other language I know of — has a single unambiguous word that precisely captures the notion of turning an opponent's attempted deception back upon them.[6] The proverbial "battle of wits" (itself coined by 1912) has contributed to common speech many words that almost but not quite qualify. To name a few: outfox (1924), outsmart (1924), outslick (1941), outwit (1943), out-psych (1974), and outsharp (1976). Finally, although the ever-popular phrase "turning the tables" precisely describes this meaning, it does so too figuratively and not in one word.[7]

As that concept is this paper's subject, we need an acceptable term. I propose two that give a close fit: TURNABOUT and DOUBLE-CROSS. At this point let's define these terms, being particularly careful to explain what they *aren't*. In other words, minimize any possible confusion or ambiguity by defining those *other* activities that are similar but not identical.

Turnabout and double-cross are close equivalents. Both imply qualities of duplicity, double-dealing, two-facedness, playback, payback, boomerang. All those terms imply a two-sided situation — the first two where an element of deception is present on at least one side; and the last three where deception is often present. However, none require that *both* sides involve deception. Only double-cross and turnabout, in their original coinage, precisely defined this reciprocal deception — although, as we'll see, both terms later drifted into ambiguousness.

---

6 How do other languages handle this problem? Systematic research might be enlightening, perhaps even help develop theory. All of the following foreign words have multiple & ambiguous meanings. The German *Dopplespiel* and *überspielen*. French *duplicité*, *duper*, & *contre-ruse*. Spanish *traiciónar* or *embustero*. Italian *inganno*. Russian *двурѲшичество*. Japanese *uragiru*. Classical Greek *tropos* and its modern derivative *trope* (but only in the sense used in linguistics). This ambiguity also seems true of the many related Chinese terms.
Contrary to my expectation, I'm unable to find any explicit term in the martial art s. For example, although the art of Japanese *jujutsu* (and its Chinese & Korean equivalents) explicitly concerns tactics of turnabout, parry, or rebound, this term translates only as "the way of softness" or, at best, "the way of yielding".

7 This phrase derives from the contemporary game called "The Tables", a medieval English form of backgammon. See *Oxford English Dictionary* (2nd Edition, 1989), under "table", *sb.* 4.c.

While comparatively rare in military & intelligence operations, the process of turnabout or double-cross has substantial precedents in other disciplines & fields. This is particularly true in the professional worlds of gambling cheats, magicians, and con artists. This should be expected as these are the main fields in which deception predominates.[8]

The raw concept of deception turnabout has long been recognized — and applauded. No doubt this is an overflow from the frequent exposure of ordinary citizens to the greed-driven enterprises of daily life — social, commercial, & political — that generate at least some degree of "street smarts". Thus back around 1595 Shakespeare, knowing his audience would understand, could write, "That is good deceit which mates him first that first intends deceit."

The concept is common enough to receive its own jargon terms in everyday language. Thus "Turnabout is fair play." And the still widely familiar phrase, "the biter bit", originated back in the late 1600s when "biter" meant a con artist. Folklore and popular fiction is awash with variations on the Faustian theme of "Deals with the Devil", particularly delightful in those rare best-outcome cases of "Beat-the-Devil".[9] The criminal underworld and professional gamblers both have "the cross" and the "double-cross". Sting operations of one sort or another long preceded the adoption & adaptation from old (by 1930) criminal slang by 1977 by American police to mean their luring of greedy criminals into a trap.

However, and most importantly for our purpose, magicians not only commonly practice this art but are the only group that has evolved a theory, which they call The Sucker Gag. This project will identify & summarize these precedents and derive the underlying principles.

Why then is double-cross such a relatively unusual response to deceptive actions in military and counterintelligence contexts? First, let's focus on military practice. Here I suggest the rarity of double-cross is a direct consequence of the general rarity of deception in contemporary Western military doctrine & practice. In other words, because modern Western military systems generally place small value on and make little use of deception, they artificially limit the possibilities for counter-deception of the type we call double-cross. This wasn't always so. Deception theory & practice has ranged from very low to very high levels from time-to-time, across different cultures, and the various

---

8   The only specialty that comes close is counter-espionage (CE). Although deception is the main tool of many other types of workers, such as sales personnel, lawyers, police, impostors, and false prophets, it isn't a *necessary* condition of their work.

9   A classic example is Stephen Vincent Benét's short story, "The Devil and Daniel Webster" (1937). English theater-goers have been exposed to plays with this turnabout theme since as early as 1639 with Robert Davenport's *A New Trick to Cheat the Devil*.

professional & academic fields.[10] Even Chinese culture, which has reached the highest levels of guilefulness during short periods in its long history, seemingly hasn't reached a threshold of deception where double-crossing has become common enough to earn an unambiguous term.[11]

2.1. Turnabout = Double-Cross

> Hitherto honest Men were kept from shuffling the Cards, because they would cast knaves out from the Company of Kings, but we would make them know, Turn about was fair Play.
>
> — Captain Dudley Bradstreet, *The Life and Uncommon Adventures* (1755), 338

The terms turnabout and double-cross are used here as synonyms. Trick the trickster was the original meaning of both.

In the case of turnabout, the above quotation from Bradstreet, dated 1755, is the earliest printed use I've found.[12] Appropriately for the audience of this paper, Capt. Bradstreet was a retired British spy who'd acted as a treacherous agent-in-place at the court of Scottish rebel Prince Charles "Bonnie Prince Charlie" Stuart.

The term double-cross or double-X has lost much but not all of its original notion — inherent in its very wording — of a doubling back or a return. This happened gradually while the fine old English word "cross" drifted into its current state of obsolescence, as explained in the next section (2.2).[13]

Thus, while double-cross — particularly in this hyphenated form — as an equivalent of turnabout remains etymologically correct, turnabout is preferable only because it's less ambiguous. Double-cross, double-crossing, and double-crosser are best used only in strictly unambiguous contexts.

---

10 Whaley, *The Prevalence of Guile: Deception through Time and across Cultures and Disciplines* (FDDC, 2007).
11 Whaley, *The Prevalence of Guile* (2007), 28-35.
12 *The Oxford Dictionary of Quotations* (Oxford: OUP,1935). Oddly, the earliest citation in the *Merriam-Webster Collegiate* (11th Edition, 2003) is only 1789.
13 I overlooked this bit of word history when writing earlier papers where I conflated "double-cross" with simple deception and consequently & mistakenly substituted "triple cross".

2.2. Cross & Double-Cross

> "You're a thief among thieves, and those who don't double-cross get crossed."
>
> — Dashiell Hammett, "The Big Knockover," *Black Mask* (Feb 1927)

To understand the double-cross requires a thorough understanding of its foundation, its pre-existing condition. And that precondition is, in a word, betrayal — for that initial betrayal sets the stage for the second betrayal. The "double cross" is always preceded by the "cross" — that now nearly obsolete slang word which originally (by 1802) meant the act of treachery of "throwing" the fight in a boxing match.[14] The very fact that a useful word like "cross" has become fairly obsolescent is evidence of our contemporary impoverished theories of deception and counter-deception, of cross and double-cross.

Cross (coined by 1802) was soon followed (by 1826) by double-cross. Again, like cross, coined in the boxing ring to mean crossing the original crosser (cheater).[15]

Fortunately betrayal is a common human behavior — common enough for most of us to have learned to recognize it in time to take appropriately *passive and defensive* countermeasures. But, also fortunately, betrayals are still rare enough in most transactions that most of us can enjoy the luxury of not having to expend much time plotting counteractions. However, whenever we're being "crossed" and the cost isn't acceptable, we'd be wise to have prepared some *active defense or even aggressive* response. That response can take the ironic form of tricking the trickster.

2.3. Deception & Counterdeception

> For 'tis the sport to have the engineer
> Hoist with his own petar.
>
> — Shakespeare, *Hamlet* (1603), Act III, Scene 4

---

14 *Oxford English Dictionary* (2nd Edition, 1989), under "cross", sb.29; J. E. Leighter (*editor*), *Random House Dictionary of American Slang*, Vol.I (1994), under "cross" & "doublecross". The antonym of "cross" was & still is "square."

15 J. E. Leighter (*editor*), *Random House Dictionary of American Slang*, Vol.I (1994), under "doublecross", where the earliest citation is 1826; *Oxford English Dictionary*, 2nd Edition (1989), under "double-cross", where earliest citation is 1834.

If deception means the act of deceiving, what does counterdeception mean? It is one word with two precise meanings. It had been coined in 1968 by William R. Harris in response to my challenge to invent a single word to mean "the detection of deception." Consequently, it officially entered the English language with that single meaning the following year with publication at MIT's Center for International Studies of my *Stratagem: Deception and Surprise in War*. But Harris had already recognized that the detection of an opponent's deception plan or operation was a necessary precondition for playing it back on them. We had that playback concept but we just didn't have a word specifically dedicated to it. So, after some toying with several unappealing terms — terms like "counter-stratagem" — Harris proposed and I concurred that we use "counterdeception" to cover both meanings.[16]

In the few cases where an unwelcome ambiguity might intrude, we felt that explicit definitions up-front would be enough. However, for readers who are more comfortable with a uniquely defining term, I have yielded in this paper and now recommend the existing and familiar ones of turnabout and double-cross.

2.4. Parthian Shots & Parting Shots

> "One other thing, Lestrade," [Sherlock Holmes] added, turning round at the door: "'Rache,' is the German for 'revenge'; so don't lose your time looking for Miss Rachel." With which Parthian shot he walked away, leaving the two rivals open-mouthed behind him.
>
> — Arthur Conan Doyle, *A Study In Scarlet* (1886), Pt.1, Ch.3

The above Sherlockian quote is a play upon words that nicely blends these two somewhat odd terms — Parthian shot & parting shot. In fact, although related, they have different meanings. A "parting shot" refers to the situation where the person breaking off or departing a conversation inflicts the "last word" by speaking a final and usually cutting or hostile remark.

In contrast, a "Partian shot" refers to the famous deceptive tactic of the Parthian nomadic horse archers — and all later horse nomads — who, feigning retreat,

---

[16] A detailed analysis of the term "counterdeception" and its relationship to "deception" and of both those terms to their parallels in "counterintelligence" and "intelligence" see Whaley, *Defining Counterdeception and Counterintelligence: Exercises in Nomenclature* (FDDC draft 29 May 2009, 25pp).

would turn in the saddle to shoot their arrows at their surprised & disorganized pursuers — often leading them straight into an ambush.

Although it's usually assumed that, because the Parthians were an ancient tribe, their military tactic inspired the term for the social tactic. In fact, the reverse is true, "parting shot" having been traced back to 1818, whereas "Parthian shot" didn't surface in print for a further 14 years.[17]

2.5. Flag & False Flag

> Very often it must have suited pirates not to be recognised as such and they no doubt made much use of false colours to gain surprise.
>
> — Timothy Wilson, Flags at Sea (1986), 46

Flag, of course, refers to the symbol of allegiance. And "false flag" (originally "false colours") first appeared as naval jargon for a vessel that disguised its allegiance by flying the identifying colors of its enemy or of a neutral. Thence, figuratively, false flag became military or intelligence tradecraft jargon applied to a situation where a person or group has been led to wrongly believe they are being recruited by their own or a third government or party. Widely used by all intelligence services, although only in the special circumstances where either the person being recruited wouldn't consider working for the real recruiter or where the recruiter wants deniability in the event of discovery that the person is a secret agent.

False flagging is deceptive, even cleverly so, but not itself double-crossing or a turnabout. It can, however, be part of a double-cross. A remarkable example was the practice of German Abwehr Lt.-Col. Giskes in 1942-43 to have the British SOE agents parachuted into Holland believe the reception parties that met them were friendly members of the anti-Nazi Resistance. Suspecting nothing, the agents freely discussed the secrets of their mission, their security arrangements, and their training & equipment in the few minutes or hours before being arrested.[18]

Similarly, of course, false flag impersonations are used by all police & law enforcement organizations whenever they run "sting" operations.

---

17 A surprisingly well-researched lexicography of both terms is in *Wikipedia*, "Parthian shot" (accessed 22 Nov '09). It shames even the great *Oxford English Dictionary* (2nd Edition).
18 Giskes (1953), 100-101; Lauwers in Giskes (1953), 176.

2.6. Blowback vs Playback

> I suddenly remembered our plans for the playback of Lauwers' transmitter.
>
> — H. J. Giskes, *London Calling North Pole* (1953), 73

Both blowback & playback are terms that sometimes involve a plan which has been turned around against the planner by an opponent. However, neither term sufficiently fits our purpose here because (with one exception) neither *requires* an opponent's intervention. Instead, both words *usually* suggest an event brought about either by nature (blowback as a fire exploding back upon us) or solely by oneself (playback by Shakespeare's image of a sapper blown up by stepping on his own land-mine).

The single exception known to me is that the German Abwehr's counter-espionage section (Abwehr IIIF) used the noun & verb-forms "play back" "playing back" to mean precisely the secret capture & "turning" of an enemy radio set and its operator and operating them back upon the enemy intelligence service.[19]

In sum, the word playback should only be used as a synonym for turnabout or double-cross when the context is clear. That will be my practice henceforward.

2.7. Bluff & Double-Bluff

> Real life consists of bluffing, of little tactics of deception, of asking yourself what is the other man going to think I mean to do.
>
> — John von Neumann defining his theory of games to Jacob Bronowski in WW II, as quoted in Bronowski's *The Ascent of Man* (1973), 432

Magicians, who are consistently the most successful practitioners of the Double-Bluff, call it the Sucker Gag, Sucker Trick, Sucker Effect, or simply a Sucker.[20] This type of trick is a double-bluff where the spectators are fed clues

---

19 Giskes (1953), 29-30, 59, 69, 73.
20 Whaley, *The Encyclopedic Dictionary of Magic*, 1584-2002 (2002), entries under "sucker".

that point to a logical but false solution of the method of working. The length of performance was about right, running only about five minutes.

The Double-Bluff or Sucker Gag is the second-most undetectable type of deception. The most undetectable category is discussed next.

2.8. Time Distortion Effects

> Ah! the clock is always slow;
> It is later than you think;
> Sadly later than you think;
> Far, far later than you think.
>
> — Robert W. Service, *Ballads of a Bohemian* (1921), 29

The most undetectable, the most impenetrable, of all types of deception is the set I call Time Distortion Effects. These have been discussed in detail elsewhere.[21] They take two main forms:

First, and by far the more common, is the It's-Later-Than-You-Think Effect. This is designed to induce people that there is no urgency, that they have plenty of time to decide or act, thereby lulling them into complacency.

Second, is the It's-Earlier-Than-You-Think Effect, which is designed to induce people to belief that time has already run out or is about to, thereby causing them to take premature action, panic, or simply and unnecessarily give up.

In itself Time Distortion isn't a form of playback. But it's mentioned here because, although I'm not aware of any real-world examples, it is in theory vulnerable to playback. This would happen if the intended victim of an It's-Later-Than-You-Think ploy would *pretend* to be lulled into complacency and *simulate* relaxing their state of alert. Meanwhile, in the absence of concrete examples, readers are invited to try plotting plausibly diabolical scenarios.

---

21 Whaley, *Deception Verification in Depth & Across Topics* (FDDC, Nov 2009), Chapter 3.7 ("Time").

3. The Three Steps to Turnabout

> **Bite the Biter:** To Rob the Rogue, Sharp the Sharper, or Cheat the Cheater.
>
> — *A New Dictionary ... of the Canting Crew, In its Several Tribes of Gypsies, Beggars, Thieves, Cheats, &c.* (London: c.1698)

To play our deceivers' tricks back upon them, we must become counter-deceivers. This process requires three steps — being deceived, detecting it, then playing it back against them. How does this process work?[22]

3.1. Deception

What do stratagemic warriors, practical jokers, con artists, artists, and magicians have in common? All invent a virtual world and invite us to visit. Wily general Odysseus famously lulled the besieged Trojans into unwarily accepting the Greek's gift of a Great Horse. Welsh physicist, Dr. R. V. Jones earned multiple reputations as "world's greatest practical joker", "father of electronic warfare", and "father of scientific intelligence". The American Gondorff brothers enticed rich gamblers into their opulent but phony bookie joint and tricked them into betting the losers in horse races that only the Gondorffs knew had just finished. Belgian painter Magritte, Dutch draftsman Escher, and many if not all other artists portray paradoxes to puzzle or inspire our minds. And magicians invite us to play a game of illusion with them to which only they know the rules. All are tricksters. Some intend to amuse, others to take unfair advantage. But all share the intent to deceive us.

We, the intended victims usually command several channels of input that we can use to verify our perceptions. Our deceivers, to create their illusionary world, must capture & control those channels of communication — the paths through which we receive all our information about events in the real world. The deceiver then proceeds through a process of planting a set of of incongruities that will induce us to build a false world picture in our minds. This false or virtual world must seem consistent according to any test we, the victims, can apply in the time available.

---

22 Although this process may seem obvious, no one had stated it explicitly until William R. Harris in the late 1960s. See particularly Harris (1972), Part III ("Methodology").

3.2. Detection

Deception is often rich in its details and apparent variety but never in its psychological essence. Consequently, because deceit is basically simple, a reasonable person might wonder whether its detection is a similarly simple process. And indeed it is, as explained in detail elsewhere.[23] Here it's enough to point out that just as deception is applied psychology, so is its detection.

3.3. Turnabout

Having detected an opposition's deception plan, what positive steps can be taken to turn it back upon them?

Although rare in military operations, there are precedents in other disciplines and fields. It can be called the double bluff. The criminal underworld and professional gamblers both have "the cross" and the "double-cross". And most importantly for our purpose, magicians not only commonly practice this art but are the only group that has distilled it into a principle they call The Sucker Gag.

This project will identify & summarize these precedents, derive the underlying principles, and give a few concrete examples.

■ ■ ■ ■ ■

The playback of a deception is simply mental jujutsu, neither more nor less. The defender, having detected the attacker's design, reverses it upon the attacker. The aggressor's action boomerangs, thereby tricking the trickster with his own trick. This playback or turnabout is the subject of the rest of this paper.

---

23 Whaley, *Textbook of Political-Military Counterdeception: Basic Principles and Methods* (FDDC: Aug 2007), Chapter 1.

4. Types of Turnabout

> The essence of his play was to encourage the "sucker" to think that he — the sucker — had an unfair advantage, and then hoist the victim with his own petard. It was a trick which, to the discredit of the human race, never failed to work.
>
> — Anonymous review of *Devol's Forty Years a Gambler on the Mississippi* in *The New York Times*, 4 Apr 1926

Deception fails in only four circumstances. These are when the target: takes no notice of the intended effect, notices but judges it irrelevant, misconstrues its intended meaning, or detects its method.

In most cases, the deception-counterdeception game ends at any of these four points. But it need not. Remedies available to *the frustrated deceiver* come under the head of what magicians call the theory of "Outs". But, conversely, here we'll describe only the types of stratagems available to *the successful detective* of deception. Indeed, the detective need only be partially successful to continue the game. Each of the above-named four circumstances gives an opportunity to turn the initial deception to his advantage, in part or in whole.

If detection succeeds, what then? What further counter-measures can we take against the deceiver? Shakespeare and La Fontaine not only told us that we can deceive the deceiver but promise that this will be both good and give us pleasure.

If the opponent's deception is discovered *and the opponent doesn't know this*, it then becomes possible for the detector to plan and carry out his own deception operation to confound the original deceiver. Their roles reverse and the target of deception becomes the deceiver, the quarry the hunter. This rare type of deception that pits one deception against another has gone by several terms including even "counterdeception", but that term was originally defined as "the detection of deception" with the implication that playback was a subordinate part or, at most, an extension. So this occasional use of "counterdeception" to mean only playback calls for disambiguation (to use the Wikipedia's useful term). Consequently, I've adopted the term "double cross" to distinguish this specific mode of counterdeception.

Having detected the deception, the detective can:

- Pretend to not notice the deception's intended effect or method. If the deceiver has feedback on this, he may be forced to return to the drawing board to design a new deception or waste more effort in reinforcing the old one.

- Pretend it is irrelevant, with the same result as above.

- Pretend to misunderstand the effect. Here, the counterdeception game begins in earnest, if the detective presents what is "misunderstood" as something the enemy doesn't want him to believe. This will not only force the enemy to drop the original deception but may even pressure him to abandon or modify his original real plan.

- Pretend to have detected more of the deception than is, in fact, the case or pretend to have not detected it at all. If the detective chooses the first option and exaggerates his achievement, the deceiver will react as in condition 3, above. However, if the detective conceals his detection, then he can set his own trap and bite the biter.

Following are the definitions and examples of the 9 types or categories of playback identified during the research. I presume this list is incomplete. For example, a separate category of Hiding Under Deep Cover seems plausible. However, this is the first attempt to develop such a typology.

4.1. Double-Ambush & Counter-Ambush

> "Shadows are lengthening and we've reached one of our phase lines after the fire fight and it smells bad — meaning it's a little bit suspicious .... Could be a amb ... [silence]."
>
> — Bernard B. Fall, 21 Feb 1967. These were the last words on this war correspondent's tape recorder before his vehicle set off a Viet Cong mine.

Ambush is one of the simplest forms of military deception. It is not just a prehistoric human trick, its origins are pre-human. It is a necessary do-or-die tactic for every hungry carnivore that's slower than its prey. Let the prey come to you. That's the basis of all ambushes. But that's only the first phase of what interests us here — the double ambush.

Variations on this double-ambush theme include the railway train ambush (used once by the IRA in County Derry)[24] and footpath ambushes by the Taliban in Afghanistan against both the Russians and Americans. Originally using lightly armed attackers, timed or remotely detonated munitions have become more common.

The now familiar practice of the roadside-bomb ambush-within-an-ambush as a terror tactic has been credited to the Provo wing of the Irish Republican Army in its operations against the British in Northern Ireland.[25] However, this tactic was only occasionally used by the IRA in the 1970s, in part due to the British countermeasure introduced in 1975 of using helicopter airlift to move most Army & SAS personnel & supplies between outposts.

The roadside IED ambush was revived in the 1st decade of the 21st century as a standard tactic against the USA and its allies first by insurgents in Iraq and later by the Taliban in Afghanistan. Unfortunately, because American strategic policy required the use of numerous large conventional military units with their continuous & voracious demand for resupply, it was impractical to use the effective countermeasure of off-road vehicles, helicopters, or light aircraft.[26] Ironically, the Coalition & NATO forces have set rules that play directly to the minimal capabilities of their various lightly armed opponents.

But let's backtrack three decades to begin our case studies with an IRA double ambush that used IEDs. Incidentally, the term "Improvised Explosive Device (IED)" was coined in the British Army in the 1970s, having been inspired by the IRA's then common use of such bombs.

CASE 4.1.1:
The Warrenpoint Double-Ambush, Northern Ireland, 1979[27]

The Warrenpoint ambush was a guerrilla assault on British Army forces by the Provisional Irish Republican Army (IRA) on 27 August 1979. It resulted in the British Army's greatest loss of life (18 killed) in a single incident during Northern Ireland's entire era of "The Troubles" (1968-1998). This operation was typical of double-ambushes that used IEDs in that it had two distinct phases:

---

24 BW interview with the planner-ambusher, former O/C Derry, Eamon J. Timoney, New York, 1979. Mr. Timoney and his small IRA team in Northern Ireland highjacked a moving freight ("goods") train and derailed it. That wasn't the target. The real target was the "breakdown train" that would be sent to fix the damage. As Timoney, an old railway man himself, knew the Irish rail line had only one of these pieces of equipment capable of handling a derailment. Destroying it would cause maximum inconvenience. They failed to destroy the relief train.

25 Conversations with Dr. J. Bowyer Bell in Dublin, summer 1979.

26 Conversation with Dr. John Arquilla, 2007.

27 As lightly edited and stripped of its extensive footnotes from *Wikipedia*, "Warrenpoint ambush" (accessed 18 Nov 2009).

Phase One — wherein the ambusher selects a section of roadway known to be a frequently traveled enemy supply route and sets a roadside bomb:

On 27 Aug 1997 in the late afternoon at 1640 hours a single 500-pound fertilizer bomb hidden under bales of straw in a lorry parked at the side of main road leading through the small town of Warrenpoint was detonated by remote control as an army convoy of a Land Rover and two four-ton trucks drove past. The explosion caught the rear truck in the convoy killing six members of 2nd Battalion, the Parachute Regiment.

After the first explosion the British soldiers, believing they were also under fire from IRA snipers, began firing across the close-by maritime border with the Republic of Ireland. This response managed only to kill an uninvolved civilian, an Englishman, and injure his cousin. There were conflicting reports of whether the soldiers had actually come under sniper fire or had mistaken the poppings of ammunition cooking off inside the burning Land Rover.

On hearing this explosion a nearby Royal Marine unit alerted the British Army of an explosion on the road and reinforcements from the Parachute Regiment were dispatched to the scene by road. A rapid reaction unit consisting of medical staff and a senior commander Lieutenant-Colonel David Blair, the commanding officer of the Queen's Own Highlanders, together with his signaler, Lance Corporal Victor MacLeod, were sent by Wessex helicopter. Col. Blair assumed command once at the site.

Phase Two — wherein the ambusher has prepared a second event for the real target, the emergency response team:

Exactly 32 minutes after the first explosion, a second bomb — a monster home-made 800-pound fertilizer device, exploded. It had been concealed in milk pails standing against the outer wall of the gate house at the opposite side of the road. IRA scouts had studied how British forces acted after similar roadside bombings and correctly assumed the soldiers would set up their Incident Command Point (ICP) in the nearest structure.

This second explosion completely destroyed the building and killed twelve soldiers — 10 from the Parachute Regiment died along with the two Queen's Own Highlanders. Parachute Regiment Major Jackson who'd arrived at the scene soon after the second explosion described seeing pieces of human remains over the area and the face of his friend, Major Fursman, still recognizable after it has been torn by the explosion from his head. Only one of Colonel Blair's epaulettes remained to identify him.

Consequences:

The attack caused major friction between the British Army and the Northern Irish counties paramilitary police, the Royal Ulster Constabulary (RUC). Lieutenant-General Sir Timothy Creasey, General Officer Commanding Northern Ireland, suggested to British Prime Minister Thatcher that internment be restored and liaison with the Republic of Ireland police be left with the military. Instead, RUC Chief Constable Sir Kenneth Newman, insisted that the conventional (since 1975) British Army practice of supplying their garrisons in South Armagh by helicopter gave the IRA too much freedom of movement.

The death of these 18 British soldiers became a significant factor in moving the British government toward accepting greater independence from the Crown for Northern Ireland. This was a notable case where a single act of terrorism had a direct & swift strategic consequence for the IRA, rather than either the usual long-term wearing down tactic of a continuous series of acts, much less the merely annoying "bee-sting" isolated "incident".

CASE 4.1.2:
A Magician Counter-Ambushes the Afghans, 1902

> Two can play tricks, and the conjurer should generally win!
>
> — Major L. H. Branson, *A Lifetime of Deception* (1953), 37

Afghanistan, weak but resilient, has replaced Vietnam as the preferred symbol of frustration & defeat for the world's mightiest legions. The quagmire of Vietnam repelled Chinese, French, and American legions as has the barren rubble of Afghanistan so far resisted the interventions of Russian, British, and American armies. But even within these strategic disappointments, invaders have enjoyed tactical victories. And even in that most formidable of Afghan tribal terrain, South Waziristan, as shown by the following case.

Born in England, Lionel Hugh Branson, learned magic in 1887 from reading Hoffmann's *Modern Magic* (1876), which was then the most recent and one of the better general textbooks of conjuring. Branson was strictly an amateur sleight-of-hand magus, except in 1913 when as "Lionel Cardac" he'd played London's Palace of Varieties for 3 weeks and also made a 25-minute conjuring demo film.

During his service with the Indian Army from 1899 until 1922 when invalided out, Branson was generally recognized as its best magician. His interesting and amusing memoirs prove that proficiency in conjuring and practical joking didn't hurt his military career. During that period he freely used magician's oblique and inverse thinking to break into a safe, solve crimes, detect malfeasance, resolve bureaucratic and personal dilemmas, succeed in two espionage assignments, and — as described now — ambush some ambushers.

Branson worked a simple but effective example of "The Biter Bit" in 1902. At that time he was a 22-year-old lieutenant posted to the 9th Bombay Infantry on the Indian border adjoining Afghanistan's southern Waziristan. There he was the only officer commanding 100 sepoys quartered in a tiny line-of-communications mud fort. A rogue band of local rebels raiding a road convoy at night had succeeded in stealing three camels. Branson formed a small party of 20 riflemen and set out to recover the King-Emperor's livestock.[28]

They'd just turned the corner of a dry river gully when two rifle shots came from the steep side above them. Normally the troops would have rushed that side — to gain better cover before returning fire. But Branson ordered only five men to that steep side where they were to then begin rapid fire, while he and the remaining 15 moved back onto the gentle slope, took up prone firing positions *facing away from the steep side* — and waited. As soon as the smaller group of soldiers on the steep side opened rapid fire, the main party of Mahsud warriors rose up behind them at the top of the gentle side expecting to fire into the backs of their enemy across the gully. Instead they saw only the small decoy party when surprised by the massed volleys from below. The Mahsuds were routed, the camels recovered, and Branson got all his small force back safely to their fort.

How had Lt. Branson managed to anticipate this unusual scenario? He explained, "About six weeks previously I had gone out to recover the remnants of a party of signallers who in similar circumstances had been nearly wiped out, and found that most of them had been shot in the back! I thought the matter out, and so on this occasion I gave the orders" stated above. He summed up, "Two can play tricks, and the conjurer should generally win!" Young Branson's escapade nicely illustrates a "prepared mind" drawing the correct conclusion from unusual facts (the dead having been shot in the back where they stood together and not while scattering in flight) and then applying his magician's oblique thinking to devise a counter-ambush.

---

28 Major L. H. Branson, *A Lifetime of Deception: Reminiscences of a Magician* (London: Hale, 1953), 36-37.

CASE 4.1.3:
Counter-Sniping, 1915-2009

> The most dangerous thing a sniper can face is another sniper because a sniper knows just what to look for.
>
> — U.S. Marine Gunnery Sergeant Paul S. Herrmann in A. Gilbert, *Stalk and Kill* (1997), 194

The smallest-scale and stealthiest military operation is sniping. Typically involving either a single sharpshooter or a two-person team of shooter & spotter, it is generally well-worth the special efforts required in training, equipment, and administration. In addition to the intelligence collected, it generally lowers enemy morale, slows their advancing patrols, and consistently yields the highest kill ratios of all types of infantry combat.

Counter-sniping is sniping's trickiest part. It requires a rare combination of high level skills — small arms, ballistics, scouting, tracking, and camouflage. And it demands a rare combination of personal psychological qualities — analytical thought, empathy, patience, and persistence.

The architecture of the short case sketches used throughout this paper can't begin to give the flavor of counter-sniping. It's a subject — like bomb disposal — that lives entirely in its details. Consequently, I'll close this subject with a short list of those few books on counter-sniping I've found insightful:

- Hesketh-Prichard, *Major* H. (1876-1920)

    Sniping in France: With Notes on the Scientific Training of Scouts, Observers, and Snipers. London: Hutchinson & Co., 1920, 268pp.

    Classic account of sniping, the first by a professional sniper. Includes, I believe, the earliest account of counter-sniping, Chapter XIII ("Wilibald the Hun", pp.164-175).

- Chandler, Roy F.; and Norman A. Chandler (1947-)

    White Feather: Carlos Hathcock USMC Scout Sniper. Jacksonville, NC: Iron Brigade Armory Publishing, 1997, 279pp. I recommend the "Sixth Printing: January 2005", which has a few updates.

    Biography of Gunnery Sergeant Carlos N. Hathcock II (1942-1999), a legendary US Marine Corps master scout sniper (with 93 confirmed kills) in Vietnam, 1966-67 & 1969.

- Plaster, Maj. John L. (1949-)

   The Ultimate Sniper: An Advanced Training Manual for Military and Police Snipers. Updated and Expanded edition, Boulder, Colorado: Paladin Press, 2006, x+573pp.

   A practical, richly detailed, and clearly explained manual of sniping. Includes much sound advice on equipment and general practice plus detailed guides to camouflage (pp.361-387), stalking (pp.389-400), mantracking (pp.415-424), counter-sniping tactics & techniques (pp.459-481), and counter-sniping in Iraq (pp.483 -501).

4.2. Double Back

I assign the old "Let's Double Back" trick to one of the higher levels of deception in all its applications in war & peace. And it's particularly effective in warfare because so few intelligence analysts ever expect much less look for it.

Doubling back takes two forms or modes. First, and much more commonly, geographical — where the prey literally & physically circles back to its start point. And second, retrospectively — where an obsolete or dead style is revived. Both models seemingly play upon the opponents expectation (preconception) that events unfold from point A, through point B, to point C, etc. Expectations seldom recognize the possibility of the target going from A to B and back to A.

CASE 4.2.1:
The Keyhole Satellite that Played Dead, 1977-78

This is a case of circling back. But with a stylish wrinkle, one where the target remains naked in direct-line-of-sight while its radioed signals (telemetry) secretly circle.

One of the cleverer ruses ever deployed by the CIA involved an effort to mislead Soviet intelligence analysts about the functioning of one type of U.S. surveillance satellite by incorporating a non-standard and therefore unexpected design feature. Beginning in 1977 the KH-11 "Keyhole" satellite's orbit carried it over the USSR, its high-resolution optical, infrared, and radar sensors designed to spot objects and movement indicating possible Soviet violations of the SALT agreement. Soviet analysts knew this but, as they could detect no radio transmissions beaming down to Earth, they concluded that this satellite, like so many others, was "dead". They discovered otherwise in March 1978 when they read an original copy of the KH-11's System Technical Manual to learn that the satellite's radios weren't even supposed to broadcast down but rather off to another satellite, which then re-broadcast Earthward.

Consequently the Soviet Army began evasive maneuvers with their ground missiles but these were quickly detected (in April) by CIA image interpreters. The FBI investigated and traced the suspected leak to Mr. William Kampiles, a recent CIA watch officer at the Operations Center. He'd flown to Athens where he'd sold his stolen copy of the KH-11 manual to a Soviet intelligence officer for the bargain basement price of $3,000.[29]

CASE 4.2.2:
The D-Day MOONSHINE Electronic Double Back, 1944

This case shows the second type of doubling back — where our opponent's preconception that our tactics & weapons will always show steady progress (the learning curve) is contradicted by our sudden & unexpected revival of an obsolete tactic or weapon. In other words, they expect us to introduce the Mark-III or Mark-IV model weapon or procedure but we momentarily trick them by using the retro Mark-I model. And "momentarily" is usually sufficient to buy a few days or hours or, in hand-to-hand combat, even a single winning "beat".

An example: On D-Day 1944 the British re-released at least one primitive Mark-I model electronic measure that the Germans had earlier foiled. This was MOONSHINE. It had been developed in March 1942 by Dr. Robert Cockburn's team at the Telecommunications Research Establishment (TRE). It was a radar "pulse repeater" designed to pick up a pulse from a German FREYA defense radar transmitter and reflect back a spread-out pulse on the same frequency that tricked the German operators into "seeing" a reflection that simulated a large formation of enemy aircraft.

That summer, MOONSHINE transmitters were fitted into the 20 obsolescent Boulton-Paul "Defiant" two-seat fighters of RAF No. 515 Squadron. They went operational on 17 August in flying a protective diversion for the US Army Air Force's first bomber raid on Europe, the one against Rouen, France. MOONSHINE initially proved its worth but was retired after a few months when the Germans figured it out and were no longer deceived.[30]

Over a year later, 6 June 1944, was D-Day. Because success was essential, the British & Americans had deliberately held back release of several of their latest model electronic warfare devises, hoping to spoof the Germans at least during the crucial first few hours and days of the invasion. The Germans expected various Mark II, Mark III, and Mark IV devises and had prepared to quickly identify and adjust to them. They did not expect to see any Mark I models. Knowing this, the British reactivated MOONSHINE (see CASE 4.2.2).

---

29   Whaley, *Textbook of Political-Military Counterdeception* (FDDC, 2007), 83; Jeffrey Richelson, *America's Secret Eyes in Space* (New York: Harper & Row, 1990), 170.
30   On the original MOONSHINE see Jones (1978), 291.

Operation MOONSHINE was now worked in tandem with Operation GLIMMER. GLIMMER was a long-standing procedure; but MOONSHINE was strictly retro. They combined to fool the Germans into believing a large convoy of slow-moving ships was crossing the Channel towards Calais — while the real amphibious invasion was about to take place 180 miles west at the Normandy peninsula. GLIMMER was the aerial deception designed to lend confirmatory depth to the MOONSHINE surface deception by a few small boats below that were moving toward the French coast.[31]

GLIMMER was carried out by the RAF's highly skilled No. 218 Squadron's dozen or so 4-engine Short Sterling III heavy bombers. These bombers dropped metallic strips called Window (Chaff) that would show up on radar and reinforce the illusion of a large convoy en route. To match the slow speed of surface ships, the bombers had to fly in precise formation in circles that crept gradually forward a dozen or so knots per circle. Success required absolute precision in flight & navigation. Moreover, the metallic strips had to be released every 4 seconds — otherwise the radar signature would no longer simulate vessels.

Below GLIMMER was Operation MOONSHINE — 12 Royal Navy motor launches crossing the Channel while towing 28 radar-reflecting balloons. The launches also carried the old MOONSHINE electronic receiver-transmitters to amplify and return the German radar signals from the 28 balloons, thereby making them simulate hundreds of slowly approaching ships and boats.

Because this combined GLIMMER/MOONSHINE electronic mirage simultaneously simulated two different but reinforcing types of radar signal, it gave "depth" to the illusion. In other words, it passed two mutually confirming tests of authenticity. That was good enough to sustain some plausibility for the brief 3½ hours duration of this spoof operation. It was sufficiently convincing that the Germans even fired on it. The overall theory of this spoof was sound, but the execution was too complex, leading to a number of glitches that degraded the intended illusion and therefore the degree of the German response.

CASE 4.2.3:
The Most Dangerous Game — in Fiction, 1924

> Things just ain't the same
> When the hunter gets captured by the game
>
> — Smokey Robinson, "The Hunter Gets Captured by the Game" (Motown Records, 1967)

---

31 On GLIMMER (including the revived MOONSHINE) see Barbier (2007), 70-71, 107-111; Jones (1978), 406; Holt (2004), 578, 822; Robbie & Stamp (1989), 169.

Richard Connell was a popular American short-story writer in the first half of the 1900s. His most famous story was first published in *Collier's Weekly* magazine in 1924. Its title was "The Most Dangerous Game".[32]

A classic short story, its premise — original then but often copied — is set in the following chilling lines — melodramatic, hokey even, but suffused with those mental shivers that explain the deathless appeal of all our favorite horror stories:

> "I wanted the ideal animal to hunt," explained the general. "So I said, 'What are the attributes of an ideal quarry?' And the answer was, of course, "It must have courage, cunning, and, above all, it must be able to reason."
>
> "But no animal can reason," objected Rainsford.
>
> "My dear fellow," said the general, "there is one that can."
>
> "But you can't mean—" gasped Rainsford.
>
> "And why not?"

Of course, this is no idle dinner chit-chat between a guest and a jaded hunter at the latter's remote lodge. Mr. Rainsford is on the tingling edge of realizing he's the chosen prey.

I attribute this tale's great appeal & memorability to the fact that its basic premise — hunter vs hunted — is turned around midway to become one of hunter vs hunter. In other words, when the victim recognizes that his best chance of survival lies in playing by his own rules and not his opponent's.

Connell's hunter explains this kind of deadly manhunting far better than Hemingway, who was at his best with rifle or tommy-gun against African lions and Caribbean sharks. Connell recognizes that man is the only animal that can play mind-games with its pursuer, the only one that can go beyond natural camouflage or mere random "jinkings" to out-think deceptively in order to hide, escape, or — in Connell's story — counter-attack.

Connell's story would inspire a whole genre of fictional versions — *Wikipedia* (by 2009) counting 12 movie clones, 13 movie spinoffs, 33 TV versions, 4 video games, and four comic books. I would add the Rambo movie version *First Blood* (1982), which set Sylvester Stallone against Brian Dennehy. I would also recommend as surprisingly effective both John Woo's *Hard Target* (1993) with Claude Van Damme vs Lance Henriksen and *Surviving the Game* (1994) with Ice-T vs Rutger Hauer.

---

32 Connell (1924). See also *Wikipedia*, "Richard Connell" and "The Most Dangerous Game" (both accessed 31 Jul 2009).

4.3. Double Exchange

> The principle [of double exchange], though elementary, may be made very effective. It needs to be used sparingly and with some intelligence. I have seen it employed so crudely as to suggest that the performer had but a low opinion of both the eye-sight and mentality of his audience.
>
> — Charles Waller, *Magical Nights at the Theatre* (1980), 183

Double Exchange is the principle of designing an object, procedure, or strategy so it can alternate between being fake and real. When this effect succeeds — that is, when the fake has put the opponent off-guard — it can be secretly traded for its real counterpart, thereby throwing the opponent into a state of double confusion.

This term, Double Exchange, is stage conjuror's jargon. It refers to a special type of theatrical Transformation where the magician (as the Transformist) secretly trades places with an assistant during a trick. This frees the magician to act outside the focused perception of the viewers. This type of ruse has been common practice in stage magic since 1901, notably by Lafayette, Blackstone Senior, Dariel Fitzkee, and Siegfried & Roy.[33]

Indeed, Dr. R. V. Jones, coming from a tradition of practical joking rather than magic, attached high value to the double exchange as a general principle:[34]

> A further touch of artistry in deception is to provide an alternative to your true intentions so valid that if your adversary detects it as a hoax, you can then switch to it as your major plan and exploit the fact that he has discounted it as a serious operation.

The Double Exchange is an enormously powerful ruse because the victim has been led to believe he still knows where the opponent is and misperceives it as posing no immediate threat. Surely, we should much prefer our enemies to be lulled into this relaxed and misinformed state of certainty than the usual security situation where we simply try to hide — thereby provoking our opponents into an active and urgent search to find us.

---

33 Whaley, *The Encyclopedic Dictionary of Magic*, 1584-2002 (2nd Edition, revised, 2002), entries for "double", "2.n. transformation", "transformist".

34 R. V. Jones, "Intelligence and Deception" (1981), 19. See also his *Reflections on Intelligence* (1989), 126.

Despite its power, Double-Exchange illusions are quite rare in military deception. Why? Partly because they don't lend themselves easily to improvisation — usually requiring lengthy advance planning and close coordination. However, I mainly attribute their rarity to simple lack of imagination on the part of our planners & intelligencers. They fail to see the value of including this type of ruse in their standard repertoire of tricks.

Let's start with a non-military example. A notable practitioner was Orson Welles, master illusionist, semi-pro magician, improvisational artiste, and world-class practical joker. He drew on all these skills to fend off both paparazzi — on those occasions when he wasn't exploiting them for publicity — or even the more intrusive bill collectors. For example, one morning in 1955 in London he wanted to escape his residence unhindered. He telephoned his scriptwriter friend Wolf Mankowitz requesting help. When Mankowitz arrived Orson explained his need for his friend to divert one of Orson's creditors' process servers who was loitering outside. "Wolfie" immediately understood his role: "Orson and I are huge. Orson really huge and me just my ordinary huge." Orson now used their similarity in height and girth to work a magician's Transposition "in life". He fitted out Wolf with his distinctive broad-brimmed hat and dark cloak and 10-inch Havana cigar and had him leave by the front door. While his disguised double led a merry chase down the street, Orson escaped out the back and caught a cab — just in time for a crucial business appointment.[35] Simple, yes. Trivial, no. Orson's double exchange was no different in principle than the world-famous "Monty's Double" ruse that reinforced the D-Day deceptions.[36]

CASE 4.3.1:
The Chinese 7th Stratagem: Make the Illusion Real, AD 755

> The Italians have a Proverb, He that deceives me
> Once, it's his Fault, but Twice, it is my fault.
>
> — Sir Anthony Weldon, *The Court and Character of King James* (1650)

Although the double-exchange principle is relatively rare in Western military culture, it is quite common in China. There it was codified in the 36 Stratagems as Stratagem No.7, "*Wu Zhong Sheng You*", which I translate as "Make the Illusion Real" or "From Nothing Make Something".[37] Compiled anonymously

---

35 Whaley: *Orson Welles: The Man Who Was Magic* (2005), Pt. IV, Chapter "Moby Shtick".
36 As conceived by British amateur magician Col. Dudley Clarke in 1944. See Whaley, *Deception Verification in Depth & across Topics* (FDDC: Nov 2009), Ch.3.3.
37 A particularly useful analysis of Stratagem #7 is Harro von Senger, *The Book of Stratagems; Tactics for Triumph and Survival* (New York: Viking, 1991), 85-108.

sometime around 1644, this is one of the best known and widely read of all Chinese texts on strategy — military and otherwise.

This 7th principle was probably inspired by Taoist philosopher Lao Tzu's somewhat counterintuitive notion that an illusion can conceal a reality. As with all other principles in their martial arts classics, the Chinese apply this one to all other aspects of human relations, particularly the political, economic, and in everyday living. One apt illustration of the 7th Stratagem is the famous Chinese story of "Straw Dummies for Soldiers";[38]

In AD 755, the illustrious T'ang Dynasty was threatened by a major rebellion. One rebel general, Ling Huchao, set siege to the city of Yongqiu, which was defended by a small garrison commanded by loyalist Commander Zhang Xun. This wily officer commanded his soldiers to make a 1,000 man-sized dummies of straw dressed in black clothing, attach them to lines, and let them slide down the outside of the city walls at twilight. Rebel General Ling, misperceiving this as a sortie by the city garrison ordered his archers to loose a hail of arrows against them. When Commander Zhang had his straw dummies drawn up with their thousands of "captured" arrows, rebel General Ling realized he'd been tricked.

Next evening, Commander Zhang had 500 real soldiers lowered down the city walls. General Ling, thinking them dummies sent to harvest more arrows, laughed in derision and made no preparations for battle. Zhang's small force struck hard & fast, set fire to the Ling's camp, killed many rebels, and scattered the rest like straw in the four winds.

CASE 4.3.2:
The Alamein MUNASSIB Switch, 1942

> Here was an expert in magic and spells.
>
> — Mure, *Master of Deception* (1980), 50

Col. Dudley Wrangel Clarke was indeed a masterful "dealer in magic and spells," having as a boy been taught the sorceror's arts by his uncle, Sidney Wrangel Clarke, the world's then leading authority on the history of stage magic. This special perspective combined with an unusual sense of humor that enabled him to approach all problems from that oblique angle needed for plotting deceptions. This made him an ideal candidate for his appointment in December 1940 to found and head A Force, the British Army's first formal deception planning-cum-operational unit. Based in Cairo, it would run all

---

38 Von Senger (1991), 88-89.

British & Allied deception in the Mediterranean & Near Eastern theaters during WW2.

Without its author having realized it, deception's culmination in October 1942 at Alamein already represented the *highest level of a coordinated deception plan*. Indeed, in most ways it even equaled that reached 19 months later by the Normandy D-Day plan (BODYGUARD).

And, in one extraordinarily innovative deception method, it actually exceeded BODYGUARD. That was Operation MUNASSIB, conducted deep in the desert on Alamein's far southern flank. Major David Mure, one of Clarke's officers explains:[39]

> Here took place a double-bluff along the lines of the struts under the wings of the real aircraft [CASE 4.4.2] and the mixing of real and bogus tanks in the [notional] 4th Armoured Brigade. The bogus guns in the pits were first given away to the enemy as dummies through the camouflage being carelessly placed on them. Just before a diversionary attack was made in the south by 7th Armoured and 44 Division, real guns were substituted.

This ruse helped produce a crucial 2-days surprise in timing of the attack at Alamein.

Double bluffs are risky — Dudley Clarke was particularly careful in deploying them. However, Major Mure used this case to illustrate a general point about how this type of double bluff also can be used to protect one's double agents by giving them a plausible out:[40]

> This attack in the south ... helped to provide a line of escape [an out] for the double agents who might otherwise fall under the suspicion of having sent deliberately misleading reports. They would, at least, retain some credibility through having correctly identified the units making up the reconstituted Eighth Army. In fact, so skilful was their operation that the battle apparently redounded to their credit with the Germans rather than the reverse.

---

39 Mure (1980), 141. See also Barkas (1952), 192, 206; Young & Stamp (1989), 73, 74; Holt (2004), 243, 829; Jones (1978), 244, 291, 406; Jones (1989), 123, 126.
40 Mure (1980), 141.

CASE 4.3.3:
The Actor Tricks the Magicians, 1922[41]

> For what is acting but lying, and what is good acting but convincing lying?
>
> — Laurence Olivier, *Confessions of an Actor* (1982), 20

A conjuror is "an actor playing the part of a magician." We might think that one actor watching another would have his critical facilities at high pitch. An instructive example of critical and detection failure is the following. In 1922, accompanied by eleven fellow-magicians, Dr. Harlan Tarbell, later a world-famed teacher of magic, attended a performance at the Times Square Theatre of *The Charlatan*, starring Frederick Tiden as Count Cagliostro, the 18th Century alchemist. They sat in the audience, amused by the melodrama and the several simple tricks used by Tiden to enliven his show, none of which would fool any magician. However, at one point, the villain, a lawyer, maneuvers the hero into a dilemma where Cagliostro must either back down in disgrace or accept a challenge by his enemy to a public test of his magical skills under conditions risking almost certain exposure. He accepts the challenge.

Cagliostro, in full view of the villain and his gang of skeptics, displays his apparatus: a handful of sand, a flowerpot of clear glass, a tall paper cone, and a seed. The lawyer seizes these items, carefully inspecting each, and shows them to his cronies (and to the audience beyond the proscenium). They are seen to be exactly what they appear to be and otherwise empty. Cagliostro proceeds to pour the sand into the pot, plant the seed in it, cover it with the cone, and step back. The lawyer again rudely intervenes to verify that he has not been tricked. Cagliostro then steps forward and confounds his enemy by raising the magic cone to reveal a full-grown flowering rosebush.

While the rest of the audience cheered Cagliostro's vindication, the twelve magicians sat amazed. They'd come expecting to watch an actor use a few obvious old tricks but perceived instead a new master magician so skilled that he had surprised even them with some great innovation in the otherwise jaded Indian Mango Tree illusion. These professional magicians simply could not detect how Tiden had done this trick without the usual recourse either to an assistant who sneaks the flowerbush to the conjuror or to some gimmicked container from which the conjuror, working alone, produces it. Some wholly new principle seemed involved. Confronted after his performance, Tiden was delighted but surprised that he had deceived a panel of experts. He frankly

---

41 Taken verbatim from Whaley, *The Maverick Detective* (manuscript).

confessed that he was *only* an actor, and cheerfully explained his "obvious" deception:

Tiden knew better than the magicians how thoroughly a theater audience can be led to suspend disbelief. So Tiden, as actor, had audaciously and blatantly chosen the enemy of the hero for his assistant. Thus, the fictitious villainous lawyer, while interrupting the second time to verify the apparatus was still empty, simultaneously, as Tiden's real fellow actor, "loaded" the empty paper cone.

Tiden's Mango variation had simply used the usual assistant but masked this fact by a psychologically deceptive twist. While most magician's assistants are openly part of the act, a special sub-type called the confederate is not. Tiden's innovation was to apply the accepted theatrical convention of actor/role to blend his actor-assistant with his role-villain, to get a psychologically invisible confederate.[42]

I trust the lesson for deception planners and deception analysts is obvious.

4.4. Hiding in Plain Sight

> The truth often lies on the surface, or close to it, and yet is no less hidden for that.
> — David Lehman, *The Perfect Murder* (2000), 84

> The devil meets with this lawyer. Says he can make the lawyer a senior partner, but the lawyer has to give him his soul and the soul of everybody in his family. The lawyer stares at the devil and asks, "So what's the catch?"
> — James Patterson, *Violets Are Blue* (2001), Ch.5

---

42 Harlan Tarbell, *Tarbell System Incorporated: Magic* (Chicago: Tarbell Systems, 1927), Lesson 16, pp.4-5. Reprinted in Harlan Tarbell, *The Tarbell Course in Magic*, V.2 (1942), 34-35. See also Henning Nelms, *Magic and Showmanship* (New York: Dover, 1969), 3. The play, which premiered on April 22nd, had been written by Ernst Pascal. Technical advisor on Tiden's magic tricks was the inventive stage magician-mechanic, Guy Jarrett, who gave a less dramatic account of this show. See Jim Steinmeyer (*editor*), *Jarrett* (Chicago: Magic Inc., 1981), 74, 80, 85. Note that both Tarbell and Nelms consistently misspell Tiden's name as "Tilden".

The doubly deceptive technique of hiding in plain view is practiced by magicians and propagandists often enough for them to give it a name. Magicians call it a "swindle". That's a trick where, paradoxically, an effect is made to seem to have occurred when, in fact, nothing did — or was so obviously self-working as to be absurd.[43] Propagandists call it the "Big Lie", which is defined along the general lines of, "If you tell a lie big enough and often enough, people will come to believe it."

The classic hiding-in-plain-sight story is "The Purloined Letter," written & published in 1844 by Edgar Allan Poe. The police ransack the apartment for the missing letter — which sits unnoticed at eye level in a letter rack. The story is so famous it provokes self-congratulatory "Can't Fool Me" assertions by literate modern police detectives. Such self-confidence is the mark of the ideal dupe, every deceiver's delight. Ordinary straight-forward thinkers like Poe's police are systematically out-thought by the kinds of oblique thinking found in all consistently successful deceivers. For example, practical jokers and magicians, as we'll see in the following two cases.

CASE 4.4.1:
Indomitable Jones and the Malta Radar

> One old adage is worth remembering here: a
> good way to hide a pebble is to put it on a beach.
> — R. V. Jones, *The Wizard War* (1989), 130

In 1941 the small island of Malta was Britain's isolated and vulnerable outpost in the mid-Mediterranean just 55 miles below the hostile coast of Sicily. Its main defense was three old biplane fighters which faced the German Luftwaffe alone. The Germans had installed powerful new radar jammers on Sicily that rendered Malta's early-warning radar useless. The local Signals Organization urgently informed the Air Ministry in London that they were now badly jammed and asked if the Ministry could help.

This question was passed to the Air Ministry's head of Scientific Intelligence, Dr. R. V. Jones. As he would recall:[44]

I knew that the Germans judged the success of their jamming by listening to our radar transmissions to see whether, for example, they ceased to scan, as they might well do if they could not be used. I therefore signalled Malta to go on scanning as though everything were normal and not to give any kind of

---

43 Whaley, *The Encyclopedic Dictionary of Magic*, 1584-2002 (2nd Edition, 2002).
44 Jones (1978), 256-257.

clue that they [the Malta radar stations] were in difficulty. After a few days the Germans switched their jammers off.

At the end of the War, I spent several days talking to General Martini, the Director General of Signals of the Luftwaffe, when he was a prisoner-of-war. He had been in his post since 1933, and had a long and detailed memory of the many events in which he and I had been opponents. At one point he specifically asked me about the jamming of Malta, and he told me that he had installed the jammers fully expecting to paralyse the Malta radar, but they seemed to have had no effect. He wanted to know what kind of anti-jamming devices we had installed in our radars so as to render them immune. He laughed ruefully when I told him that he had in fact succeeded, but that I knew the clues on which he would judge his own success, and had therefore advised the Malta radars to pretend that they were still working.

CASE 4.4.2:
Dudley Clarke and the Aircraft Dummies

Colonel Dudley Clarke headed "A Force", the discretely cover-named British military deception planning and executive organization in the Middle East Command and Mediterranean Theatre during World War II. That he was a competent amateur magician, having practiced that basic art of deception since age 12, may explain in large part his extraordinary ability to solve problems by oblique thinking. The following exchange is a perfect example of magician's thinking. Here we see Col. Clarke turning an enemy's discovery of one of his camouflage deceptions into a counterdeception ploy — one instantly conceived and quickly carried out.[45]

Major Oliver Thynne had joined Clarke's "A" Force deception team in Cairo in the spring of 1942. Soon afterwards he discovered from Intelligence that the German aerial observers had learned to distinguish the dummy British aircraft from the real ones because the flimsy dummies were supported by struts under their wings. When Major Thynne reported this to his boss, Colonel Clarke, the "master of deception", fired back, "Well, what have you done about it?" To which the novice deceptionist fumbled, "Done about it, Dudley? What could I do about it?"

When I posed this problem as a snap test question to my classes of American and foreign Special Operations officer-students in the period 2003-2005, less than one in ten could anticipate Clarke's solution, which was: "Tell them to put struts under the wings of all the real ones, of course!"

Of course? Hardly. A commander with a straightforward mind, having recognized a telltale flaw in the dummies, would have ordered the camouflage

---

45   Mure (1980), 98, 141.

department to correct it. But Clarke's devious mind instantly saw a way to capitalize on the flaw. By putting dummy struts on the real planes while grounded, enemy pilots would avoid them as targets for strafing and bombing.

Additionally — although I'm not aware this was tried — a judicious mix of real, dummy, and invisible dummy struts might cause the German photo-interpreters to both mislocate the real RAF planes and underestimate their numbers.

4.5. Hiding under Thin Cover

> The most effective way to conceal a simple mystery is behind another mystery. This is literary legerdemain. You do not fool the reader by hiding clues or faking character à la [Agatha] Christie but by making him solve the wrong problem.
>
> — Raymond Chandler, "Twelve Notes on the Detective Story, Addenda" (1948)

The classic example, the most famous of all military deceptions, is the Great Horse of Troy, the legendary gambit devised by the Greek's most cunning warrior, Odysseus, to smuggle troops under cover into the besieged Trojan capitol of Troy. That example has inspired many imitations. Three other examples follow:

CASE 4.5.1:
The FAREWELL Gambit, 1982-84

> Why not help the Soviets with their shopping? Now that we know what they want, we can help them get it.
>
> — Gus Weiss, draft manuscript, as quoted in
> Reed (2004), 267-268

Beginning in 1970 Soviet intelligence services (KGB & GRU) conducted a large-scale and largely successful espionage effort (called Line X) to steal the latest Western proprietary & secret technology. American intelligence had been given the full details of these Soviet technological intelligence (TECHINT) efforts by the French. They'd obtained this information through their volunteer agent-in-place in Moscow, KGB Col. Vladimir Vetrov (code named FAREWELL). From 1981 until his arrest in 1982 (and execution in 1983) Vetrov had been feeding French intelligence some 4,000 secret documents that identified the 200+ KGB TECHINT espionage officers operating abroad under diplomatic cover plus leads to 100+ of their locally recruited agents. These efforts are well-documented and widely known. Less known was a cunning American countermeasure that played the KGB's own game back upon them.

This counter-deception was devised by Gus Weiss, a senior advisor to the NSC under US President Reagan. Dr. Weiss (he held a 1966 NYU PhD in Business Administration) had been a senior advisor to US Presidents Nixon, Ford, and Carter. In Jan 1982, he presented his scheme to CIA Director Casey:[46]

> I proposed using the Farewell material to feed or play back the products sought by Line X, but these would come from our own sources and would have been "improved," that is, designed so that on arrival in the Soviet Union they would appear genuine but would later fail. US intelligence would match Line X requirements supplied through Vetrov with our version of those items, ones that would hardly meet the expectations of that vast Soviet apparatus deployed to collect them.
>
> If some double agent told the KGB the Americans were alert to Line X and were interfering with their collection by subverting, if not sabotaging, the effort, I believed the United States still could not lose. The Soviets, being a suspicious lot, would be likely to question and reject everything Line X collected. If so, this would be a rarity in the world of espionage, an operation that would succeed even if compromised.

Casey liked the proposal and President Reagan was enthusiastic. So Director Casey deployed a coordinated CIA/FBI/DOD operation that systematically planted false technological data and even sabotaged gadgets on the KGB & GRU spies.[47]

Both Gus Weiss and NSC member Thomas Reed have credited this sabotage program with one spectacular result. This involved computer software rigged with a "Trojan Horse" that the Russians installed as a component in control

---

46 Gus W. Weiss, "Duping the Soviets: The Farewell Dossier," *Studies in Intelligence*, Vol.39, No.5 (1996).
47 Weiss (1996), 121-126; Thomas C. Reed, *At the Abyss: An Insider's History of the Cold War* (New York: Presidio Press, 2004), 266-270; *Wikipedia*, "Farewell Dossier","Gus Weiss" & "Vladimir Vetrov" (all accessed 6 Dec 2009).

stations along their new natural gas pipeline from Siberia to Europe. This pipeline was a major part of Soviet Premier Gorbachev's design for strategic economic influence in Eastern and Central Europe. When the Trojan Horse activated in June 1982, it caused an enormous (3-kiloton) explosion in the pipeline. As Reed explained:[48]

> In order to disrupt the Soviet gas supply, its hard currency earnings from the West, and the internal Russian economy, the pipeline software that was to run the pumps, turbines, and valves was programmed to go haywire, after a decent interval, to reset pump speeds and valve settings to produce pressures far beyond those acceptable to the pipeline joints and welds. The result was the most monumental non-nuclear explosion and fire ever seen from space. [The explosion was, in fact, so large that] "at the White House, we received warning from our infrared satellites of some bizarre event out in the middle of Soviet nowhere. NORAD feared a missile liftoff from a place where no rockets were known to be based. Or perhaps it was the detonation of a small nuclear device. ... Before these conflicting indicators could turn into an international crisis, Gus Weiss came down the hall to tell his fellow NSC staffers not to worry.

Fortunately, no one had died. In 1984 Director Casey ordered the Soviet's Line X network rolled up. By then he was satisfied that this project had run its course, having no doubt become known to Soviet intelligence through Vetrov or other sources. Accordingly, America's NATO allies were notified and over 200 members of this Soviet network, including 47 in France alone, were expelled from their diplomatic posts.[49]

CASE 4.5.2:
HEINRICH Helps Set the Stage for the Battle of the Bulge, 1944

Here is a case where the Trojan Horse was pure information. It was November 1944, during the secret preparations for Hitler's last great counterattack to throw back the Allied invasion of the German heartland that, driving into the Ardennes Forest, would become known as the Battle of the Bulge. Abwehr Lt. Col. Hermann Giskes was one of the German Army's most creative deception planners, as he's already proved (CASE 4.8.2). On September 9th he'd been transferred from Holland to German Army Group B defending the German frontier. His contribution to this battle was to distract Allied Intelligence with a

---

48 Reed (2004), 268-269.
49 *Wikipedia*, "Siberian pipeline sabotage" (accessed 9 Dec 2009); William Safire, "The Farewell Dossier," *New York Times*, 2 Feb 2004.

diversion. Accordingly he devised Operation HEINRICH, which would make it seem the German army was preparing a local two-prong assault to recapture Aachen, which was well north of the Ardennes.

Giskes's scheme was elegant and diabolical. First he recruited a German Saxon engineer friend who ran a local slave-labor camp with mostly anti-Nazi Belgians & Luxembourgers. Then, and after some false starts, the Saxon recruited one of these workers by pretending he was a secret Communist who would help him escape to the nearby American lines. He primed the man with a few written details about the forthcoming (notional) German attack on Aachen. Additionally, he was to tell the Americans that, if they wanted further military details and updates, to broadcast a coded message — "And tonight we send regards to Otto from Saxony" — on the nightly German-language news broadcast from the Allied-run "black" radio station in Luxembourg City. Ten evenings later Giskes was pleased to hear this message. He then had his Saxon friend recruit and release 10 more prisoners, each primed unwittingly with more false intelligence. Their information reinforced the Allied preconceptions of some sort of German move against Aachen on the eve of their real massive push into the Ardennes.[50]

Col. Giskes's plan had involved three classic types of deception (including one double-cross). First, its false-flag recruitment of the slave-laborers. Second, its deployment of these recruits as, to use Sun Tzu's explicit terms, "expendable" or "dead" spies. And third by exploiting them as disinformation-loaded Trojan Horses.

CASE 4.5.3:
The Bar Code Orange Con, 2003-2009

> When Mississippi riverboat gambler William "Canada Bill" Jones was warned by a friend that the card game he was playing in was crooked, he said, "I know it is, but it's the only game in town."
>
> — Traditional

---

50 Charles Whiting, *Ardennes; The Secret War* (New York: Stein and Day, 1985), 84-90; and, for greater detail, Charles Whiting, *Ghost Front: The Ardennes before the Battle of the Bulge* (Cambridge, Mass.: Da Capo Press, 2002), 102-107, 113-115, 203. Important only for the author's personal interviews with Hermann Giskes, Hans Habe, and Otto Skorzeny.

> As all con artists know so well, the success of any scheme depends entirely on their ability to suspend — if even for an instant — the ability of their intended victim to think.
>
> — Lieutenant Dennis Marlock, "How to Con a Con," *FBI Law Enforcement Bulletin* (Jul 1992), 2

On 21 Dec 2003 — a financially & publicly inconvenient four days before Xmas — the U.S. Department of Homeland Security (DHS) announced that the terrorism attack threat level was being raised from YELLOW ("Elevated") to ORANGE ("High").[51] Secretary Ridge explained this was in response to intelligence from "credible sources" indicating "that extremists abroad are anticipating near-term attacks that they believe will either rival, or exceed, the attacks" of 9/11.[52]

Fast forward six years to December 2009 when Playboy magazine's website published a long & detailed feature story titled "The Man Who Conned the Pentagon." It claimed to give the back-story to this alert. The author was Aram "Sonny" Roston, a 43-year-old American freelance investigative journalist then regularly contributing to the The Nation weekly and to GQ. Previously he'd been a police reporter in New York City, a foreign correspondent, a contributor to Washington Monthly, a CNN News reporter, NBC Nightly News producer (including for a 2005 Emmy Award-nomination), and author of a generally well-received biographical exposé of former Iraqi exile leader Ahmad Chalabi, *The Man Who Pushed America to War*, published in 2008 by Nation Books. All left-liberal credentials. Hence a plausible bias to explain the initially small but damaging anti-government slant that then colored all follow-up stories for the next few weeks.

After Roston broke this story, the great majority of American news media reports were gleefully negative throughout that month. Moreover, this negative coverage spanned the full range of political bias from The Playboy Forum (Internet), The Rachel Maddow Show (TV), and Amy Goodman's Democracy Now! (radio) on the left to The Wall Street Journal and Fox News on the right.

---

51 The 5-color threat levels, introduced in 2002, were GREEN (Low), BLUE (Guarded), YELLOW (Elevated), ORANGE (High), & RED (Severe). Levels were set by a majority vote of the President's Homeland Security Advisory Council with the President's concurrence. At the time of the Xmas Orange alert the Council comprised DHS Secretary Ridge, Attorney General Ashcroft, FBI Director Mueller, CIA Director Tenet, SecDef Rumsfeld, & Secretary of State Powell.

52 "Remarks by Secretary of Homeland Security," Department of Homeland Security press release, 21 Dec 2003.

There is cause for embarrassment and much blame to share for the false alarm that triggered the artist only fully disclosed & acknowledged six years later in 2009. It certainly shouldn't — and probably needn't — have happened. It had been based entirely on the Department of Homeland Security having credited faked information. Moreover, this had not been a deception put over on the DHS by some malicious and cunning foreign intelligence service but a confidence scam put over by a greedy American snake oil salesman.

Two things must be said at the outset. The U.S. intelligence community (IC) had performed substantially better than implied by the initial December 2009 press reports. And it also showed that many American journalists were operating under the old "rule" of sensationalism that they should never let facts stand in the way of telling a good story.

The American media had consistently implied or reported three false claims: 1) That the IC had not just been momentarily suckered by the contractor's faked claims but that it somehow took great investigative prowess by the American news media to bring out the correct facts; 2) as recently as 2009 the "Pentagon" (specifically, the US Air Force) was still blindly doing business with the fraudster; and 3) that this fraudulent intelligence was the only indicator of some imminent attack in the USA. The actual facts are:

> **First**, the IC (particularly the CIA & FBI) became fairly quickly suspicious and eventually (within a year and a half) undeceived. Roston himself gave proper credit to the CIA; but by burying it one-third into his long article, he effectively let hasty media news reporters miss this point. What he said in paragraph 22 was:
>
>> A former CIA official went through the scenario with me and explained why sanity finally won out. First, Montgomery never explained how he was finding and interpreting the bar codes. How could one scientist find the codes when no one else could? More implausibly, the scheme required Al Jazeera's complicity. At the very least, a technician at the network would have to inject the codes into video broadcasts, and every terrorist operative would need some sort of decoding device. What would be the advantage of this method of transmission?
>
> **Second**, the U.S. Air Force had stopped well before early 2009 being suckered by Montgomery's already discredited scheme but even makes the credible claim that it was running a sting against the fraud.
>
> **Third**, as DHS Secretary Ridge recalled, there had been "a significant increase" in other threat indicators, although he is clear that it was the

"unique source" disclosed later by Roston that triggered Code Orange. In the Secretary's words:[53]

> We had been monitoring a unique source of intelligence and communication that seemed to identify specific flights originating from Great Britain, France, and Mexico that were either targets or carried terrorists. This coupled with other information resulted in discussions for several days in our meetings. One morning, after everyone had their say, the president, in a conversational and commonsense manner asked, "Which one of you, based on this information, would put your family on one of those flights?" No one said they would. The president had made his point. The flights would be cancelled.

This story illustrates a basic principle of information or psychological operations. Namely, that transparency trumps secrecy in all cases where the cost of letting a third party — friend or enemy — be first to reveal an embarrassing story. In other words, always stay ahead of any potentially embarrassing story by breaking it yourself. Open disclosure of all key facts in a case is the most potent way to preempt idle or even hostile speculation of the sort that gives "legs" to rumor.

This isn't the usual case where some official spokesperson puts a Madison Avenue "spin" on a story, "shades the truth", or is being "economical with the truth" (as the British government discovered to its embarrassment in 1986 in the Crown vs Peter Wright case.[54] Quite the contrary. This calls for the unvarnished truth. The SOP precedent for such openness is the "Corrections" slot on our newspaper's index page — unobtrusive and, where warranted, a quite expression of apologetic "we regret". Of course, once transparency is accepted as the wiser course, the only tactical decision calling for a measure of cunning is who makes or where do you place the correction. In this case a statement with handout during a routine press conference by the Department of Homeland Security might have been appropriate. Had such a policy been in place since 2004, the U.S. government might have avoided the flood of adverse press coverage that its secrecy eventually invited.

■ ■ ■ ■ ■

---

53 Tom Ridge, *The Test of Our Times: America Under Siege* (New York: St. Martin's Press, September 2009), 205.
54 Malcolm Turnbull, *The Spycatcher Trial* (Topsfield, MA: Salem House, 1989), 108, plus analysis in Whaley, *The Maverick Detective* (manuscript).

The previous paragraphs have described the consequences — media frenzy & government embarrassment — of Roston's exposé. It's time to correct the back-story. How did some agencies of the U.S. government get conned? And for how long? Is it reasonable to expect that intelligence or deception analysts would have given timely warning?

Dennis Montgomery could be called The Bernie Madoff of the Codebreakers. Both were con artists, both used the same M.O. of smoke & mirrors, and both were selling fictitiously high-value information. Madoff pretended to be a financial wizard selling an inside track to the stock market.[55] Montgomery pretended he'd broken an Al-Qaeda cipher — one used to give targeting orders to its terrorist cells in the field. These, he claimed, were "bar codes" transmitted over Al Jeezra TV, cleverly hidden in visual TV images, that gave the GPS coordinates and airline flight numbers of Al-Qaeda's next target.

This is still a developing story, one with more allegations, innuendo, conspiratorial speculation, and political biases than solid evidence, reasoned skepticism, tight analysis, and broadly agreed conclusions. Good! That makes it a textbook challenge for the deception analyst.[56] Moreover, this story illustrates the chief difference between the traditional (i.e., pre-WW2) historian and the intelligence analyst. Both seek to explain human events; but nearly all traditional historians examine only past events, while intelligence analysts commonly cover current ones while they're still unfolding.

This story requires viewing with cautious skepticism because, so far, it is not only single-sourced but from a source biased toward conspiracy theories, particularly ones that can embarrass the U.S. government. Consequently, this case of the Xmas 2003 Code Orange terror alert will cite without qualification only those facts alleged by Roston that are supported by at least one independent & reliable source. I stress my own finding that Roston did a generally fair job of fact-checking and reporting, which has been often garbled & sensationalized in the American & British press, including even Fox News.

So, what of Roston's claim that Dennis Montgomery had run a confidence game? If true, Montgomery was, like Bernie Madoff, faithful to the rules of behavior by which all confidence games are played. Consider the following list of those rules. They are the typical con artist's M.O. See the yellow and red flags that each created; and remember, these are all clues that were available to the

---

55 For the red flags on Madoff see Housworth (2008); Arvedlund (2009); and Whaley, *Deception Verification in Depth & across Topics* (FDDC, Nov 2009).

56 Similar challenges to deception analysis have been accepted and their authors' conclusions largely proved correct in the two different models of deception analysis developed by both Ladislav Bittman in his *The KGB and Soviet Disinformation: An Insider's View* (Washington, D.C.: Pergamon-Brassey's, 1985) and Richards J. Heuer Jr in his "Nosenko: Five Paths to Judgment," *Studies in Intelligence*, Vol.31, No.3 (Fall 1987), 71-101. Bittman's model reverse engineers hypotheses based on his experienced intuition. Heuer's model is a winnowing down of alternative hypotheses through a rigorous application of abductive logic.

victims while the con was still being played out. No hindsight, 20/20 vision, or Monday morning quarter-backing was needed to see enough warning signs to at least motivate the victim to probe below the surface of these deals that were truly too good to be true.

RULE 1:

You are selling only one product — yourself. Not wealth, fame, health, immortality, love, secret information — you don't even have them to sell. Just yourself. The only thing you are selling is confidence in yourself. That's why we call you a *confidence* artist. But keep it subtle — cognitive psychologists show it helps us deceivers to have a measure of self-deception, to believe our own lies. Don't worry, if such self-confidence doesn't come naturally, it will come with practice.

Both Madoff and Montgomery fit this pattern, but nothing suspicious here

RULE 1a:

If they are not complete sociopaths, play upon any moralistic, altruistic, or guilt feeling they may have. Try to convey *your* confident trust in *them* — that by helping themselves, they're also doing good by helping you. Remember, a few suckers are actually moralistic or altruistic in addition to or even instead of being greedy.

Both Madoff and Montgomery worked these vulnerabilities. This is at most just rare enough to be mildly suspicious.

RULE 2:

Know your mark and what the sucker wants. Tailor your persona and sales pitch to fit the preconceptions of each target.

Madoff knew they were greedy for money, money that could be made by investing in financial markets.

Montgomery knew his targets greedy for an algorithm, computer software that would decrypt the alleged secret messages in an enemy's alleged communication channel.

RULE 3:

Never look like you've sought out the mark. Make first contact either seem accidental or through a reliable third party — either a confederate ot, better, another dupe.

Madoff depended heavily on word-of-mouth, his reputation always preceding him.

Montgomery also worked the third-party ploy. By most accounts he was a bit too quirky to automatically inspire confidence. Consequently he let his bosses or partners speak for him, be his "stalking horse".

Specifically, after moving from California to Reno, Nevada, in Sep 1998 Montgomery co-founded eTrepped Technologies LLC, a software firm in Reno, in partnership with Warren Trepp, a wealthy businessman. Trepp put in most of the capital; Montgomery ran the operation as Chief Technical Officer and was paid $1.3 million for his software plus 50% ownership.[57]

Trepp then "sold" Montgomery to a local U.S. congressman who helped their eTreppid Technologies get federal contracts.

RULE 4:

To capture the attention of the mark, the smart con pushes all the right buttons — a deal almost too good to be true!

Madoff was a virtuoso at playing the buttons — money, money, & more money. The old too-good-to-be-true YELLOW FLAG.

Montgomery too. He claimed Al-Jazeera was regularly broadcasting coded instructions to Al-Qaeda terrorists cells in the USA. These messages were in the form of bar codes that when deciphered gave the geographical coordinates of targets in the USA and, sometimes, even the flight numbers of passenger aircraft. Exactly what the post 9/11 American intelligence services wanted! YELLOW FLAG.

Roston cites his interview with Frances Townsend, a lawyer & intelligence specialist who in 2003 had been Deputy National Security Advisor for Countering Terrorism. He quotes her as saying, "It didn't seem beyond the realm of possibility. We were relying on technical people to tell us whether or not it was feasible."[58]

---

57 The six most detailed stories preceding Roston's are John R. Wilke "Congress's Favors for Friend Include Help in Secret Budget," *The Wall Street Journal* (1 Nov 2006); William R. Levesque, "Ther's no way to evaluate eTreppid's work for SOCom," *St. Petersburg Times* (8 Mar 2007); David Kihara, "Attorney subpoened in Gibbons enquiry," *Las Vegas Review-Journal*, 1 Nov 2007; Anthony Effinger, "Yellowstone Club Divorcee Entangled in Terrorist Software Suits," *Bloomberg News*, 29 Aug 2008; Jonathan Weber, "Yellowstone Club Chronicles: The Edna Blixseth Bankruptcy," *New West Bozeman*, 11 May 2009; and David Kihara, "True Believers: Nevada company's troubles entangle Gibbons, [and U.S.] federal government," *Las Vegas Review-Journal*, 7 Jun 2009. All six accounts were retrieved from the Internet. Taken together they give every key fact in Roston except the Orange alert.

58 That Roston personally interviewed Townsend isn't clear in his *Playboy* piece but he confirms it in his recorded interview with Amy Goodman, "The Man Who Conned the Pentagon", *Democracy Now!*, 28 Dec 2009.

RULE 5:

Present your credentials. One or at most two genuine & easily verifiable ones should suffice, all others need only be plausible. Avoid seemingly obvious incongruities — although you'd be surprised how many you can usually get away with.

Madoff played his part well but he did create the two unnecessarily glaring incongruities that finally raised enough suspicion to lead to the detective work needed to expose him. TWO RED FLAGS [59]

Montgomery's scientific credentials were weak — only an AA (two-year college) degree as a biomedical technician in the early 1970s from a third rate college. But Montgomery's senior position with an up-and-coming software company (eTreppid) implied solid on-the-job competence, perhaps even genius. At most a YELLOW FLAG.

However his partner (Trepp) in the company had been the chief trader for one of the most celebrated American fiscal felons of the 1980s-90s, Michael "The Junk Bond King" Milken. Although never criminally charged himself, Trepp contributed $19 million to a settlement. Guilt by association? Maybe not but at least a RED FLAG.

RULE 6:

Never act eager to do business. Make the mark anxious to do business with you.

Madoff even refused to take money from enough potential clients to gain the useful reputation that he really didn't need anyone's money — he was doing you a favor. That's why most of his suckers believed he was a personal friend. Perhaps a pale yellow flag.

While Montgomery made it clear that he sought contracts for the company he worked for, he did play the friendly and cooperative patriot. NO FLAG.

RULE 7 (optional):

Cultivate an air of mystery. Make the sucker a willing part of a conspiracy. Be *openly* secretive ("I can't tell you."or "It's a secret.") about such things as the details of who & where you got the information (inside trading, the location of a high-value terrorist, etc) or the goods (a lost Rembrandt, the map to a secret treasure, plans for a perpetual motion engine, a proprietary formula, etc).

---

[59] Specifically, his financial reports to his clients showed them earning an unprecedentedly consistent statistical level of profit and he used an auditing firm with unusually weak credentials for the volume of transactions handled.

Madoff consistently refused to reveal any details of his money making machine — allegedly not even to his several grossly negligent wholesalers. A big bright RED FLAG.

Montgomery consistently refused to reveal his algorithm — not even to his partners. A big bright RED FLAG.

RULE 7a:

But don't overdo the Man-of-Mystery persona. Some eccentricity may be a plus but, depending on your target audience, avoid appearing unstable or crazy, i.e., don't tell them about your alien abduction experience unless your suckers are UFO buffs. You may act superior but only within "normal" range.

Madoff stuck to this rule. No flag.

Montgomery struck some employees as distinctly oddball, particularly his story of alien abduction. YELLOW FLAG.

RULE 7b:

Also avoid any appearance of incompetence or of being out of control, as demonstrated by such things as being involved in public scandals or numerous litigations.

Madoff stuck to this rule. No flag.

Montgomery had a well-publicized background of both a fiscal scandal and litigation. At least one bright YELLOW FLAG.

RULE 8:

Once they buy in, up the ante.

Madoff did this with all his clients, enticing them into increasing their bet (without at least partly cashing in) to the point that many ended giving him their life savings. YELLOW FLAG.

Montgomery, through his company, asked for further contacts for the software, which still hadn't proven itself. YELLOW FLAG.

RULE 9:

Monitor feedback. Sooner or later someone — rarely the mark but usually a curious third party such as a journalist — will wise up and blow the whistle. So be prepared to answer awkward questions. To make detection & exposure difficult, buy time with smoke & mirrors. Act offended — "Who would you rather believe, me or some disgruntled

former employee?" Or, "Oh, you'd trust your vague *intuition* against all those other satisfied clients?"

Madoff paid off some clients, made ever more promises, and became increasingly less open about his dealings. YELLOW FLAG.

Montgomery seemingly just stonewalled, gambling that time would be on his side long enough to profit. YELLOW FLAG.

RULE 10:
Know when to jump. The right time is either when a) the well has run dry, or b) the heat is on. Then get out fast. And, ideally, get out clean. That means leaving the marks too embarrassed or too compromised to complain — either too involved in an illegal deal like inside trading or too protective of their own secrets to risk their exposure in open court. Commercial firms and secret government agencies are traditionally vulnerable on this last point. Con artists call their getaway tactics the "blow off" and magicians, including Col. Dudley Clarke, call it an "Out" or an "escape".

Madoff overlooked this escape rule and is paying the price — seizure of his ill-gotten gains and a 150-year prison sentence.

Montgomery seemingly has been just careful enough to narrowly dodge prosecution on fraud charges, although he later managed to blunder into costly litigation in other questionable commercial dealings, including a catastrophic legal settlement in his battle with one large-scale bankruptcy.

Aren't these 10 rules of the con too obvious to be worth mentioning? Obviously not, otherwise there would be no charlatans, no double dealers, no double agents, and no counter-espionage officers.

■ ■ ■ ■ ■

The Montgomery case requires some brief preliminary analysis. It's clear in hindsight that his algorithm didn't work, indeed probably didn't even exist. But should anyone have been conned in at the time? Well, not any longer than it would take them to talk to a computer scientist specialized in type of filtering that Montgomery claimed for his product. When they realized he lacked both the basic education and practical experience to have plausibly invented such a revolutionary software breakthrough, expert efforts were then made to replicate his findings. When those failed it was a fair bet that Montgomery's secret algorithm didn't work, wouldn't work, and — judging from the testimony of three computer specials who'd worked under him, probably didn't even exist.

■ ■ ■ ■ ■

Montgomery's project managed to survive long enough to cause the 2003 Xmas alert but then quickly dissolved. For anyone with lingering doubts that he may have gotten a bad press, consider the timeline of the following post Xmas 2003 points:

2004, Jul	eTreppid lands a 5-year $30 million no-bid contract from the Special Operations Command (SOCom). Aerospace Daily ranks eTreppid the 16th largest defense contractor of that year.
2005, Dec	Montgomery is alleged by Trepp to have stolen "source codes" from eTreppid. M is formally fired effective 20 Jan 2006.
	Trepp sues Montgomery who counter-sues over ownership of eTreppid's military software. (See Rule 7b)
2006	Trepp told the Air Force Office of Special Investigations that Montgomery borrowed $1.3+ million, including $300,000 to cover casino debts. (Note the irony that many successful con artists lose their fortunes gambling.)
2006, Feb	During the Nevada gubernatorial campaign M accuses Nevada Governor Gibbons of having taken bribes from eTreppid for helping get military contract. The Department of Justice later cleared the governor of these charges. (See Rule 7b)
2006, 1 Mar	The FBI serve a search warrant on M's house, seizing computer disks, a laptop, and some illegal drugs. Two days later they execute a second warrant on a Montgomery storage garage, seizing more computers & discs. A judge, citing improper search, orders property returned.
2006, By Mar	Super-wealthy Edra Blixseth is so impressed by M's software claims that she founds Blxware LLC of Bellevue, Washington, and hires him as the company's Chief Scientist. In Dec she begins divorce proceedings with her Forbes 400 husband Tim Blixseth ($1.3 billion).
2007, Jul-Sep	Michael Flynn withdraws as M's lawyer on grounds of non-payment. Accuses Montgomery in LA Superior Court of being a "pathological liar" and that "He is a perfect example of the principle that if you tell a big enough lie, people will believe you." According to a 1 Nov 2007 story in the *Las Vegas Review-Journal* "Flynn states that the software at the center of theTrepid dispute is junk." The newspaper added "That software is reportedly used in the government's war on terrorism. It allows the military to search for people or objects in video images taken over battlefields from aircraft such as Predators."

2008, Sep   eTreppid & Blxware announce a confidential legal settlement whereby Blxware receives M's software in exchange for some percentage of sales. Trepp also received some compensation for M's slanders.

2008, Dec   A U.S. District Court in Nevada orders that Montgomery & Blixseth pay $25+ million as a result of the eTreppid & Blxware litigation.[60]

2009, Feb   U.S. Air Force pays $2 million to Blxware which then pays $600,000 to Montgomery.

2009, 2 Apr   A U.S. Magistrate Judge of Nevada District Court fines Montgomery $61,323 in sanctions for having perjured himself in a declaration to the court about his knowledge of the qualifications of his first lawyer (Michael Flynn) in the Trepp vs Montgomery case.[61]

2009, 16 Jul   Montgomery arrested in California on a Nevada warrant alleging he'd passed $1 million worth of bad checks in a Las Vegas casino.

2009, By Dec   Edra Blixseth, in two major bankruptcies, has most of her property seized, her Yellowstone Club liquidating for $115 million, a quarter of its peak value.

■ ■ ■ ■ ■

To conclude:

Madoff had earned 3 red flags & 3 yellow ones; Montgomery 3 red flags and 5 yellow ones in the early stages of their operations. And in Montgomery's case all before the Homeland Security's Xmas 2003 Orange alert. Those warning flags should have been more than enough to have raised suspicions. Sufficient enough that their confidence scams should have been probed & verified early on. When such vast sums are involved, cross-checks by top private investigators & forensic accountants are a prudent investment. And, in Montgomery's case, even a quick-and-dirty Google name check should have given pause.

Montgomery enjoyed substantial although considerably less success than Madoff. The intelligence community was slow but it caught on early enough to limit his damage to the Xmas surprise and no more than US$ 3 million out of a potential loss of at least US$30 million. As of New Year's Day 2010, the news

---

60   Jessica Mayrer, "Warrent approved for Edra Blixseth," *Bozeman Daily Chronicle*, 19 Feb 2009.
61   *The National Law Journal's L.A. Legal Pad*, 7 Apr 2009.

media have still put too much blame upon the US intelligence community. We are overdue for a critical reassessment.

4.6. Provoking Truth

One form of turnabout is provocation. But only when used to elicit the truth behind a lie and not in its more usual sense of simply pushing an opponent into a confrontation. Again, English has no single word for this, so let's use the phrase "provoking truth."

The double-cross here is to get liars or deceivers who are already engaging in some form of imposture to say or act out of character. In other words, induce them to break cover by revealing something that the person they claim to be would not know or the role they pretend to occupy would be incapable of doing. For example, a senior officer POW who pretends to be a lowly subaltern to avoid close interrogation but fails to salute the nominal senior POW officer. Or cryptologist Leo Marks's uncle who successfully pretended to be deaf to avoid military service until on leaving his examination he glanced at his wristwatch when one clever examiner behind him quietly asked, "Got the time on you, Guv?"[62] Mr. Marks would apply this family principle in WW2 to trick Germany's most deceptive intelligencer, as seen in the next case.

CASE 4.6.1:
Plan GISKES: Mr. Marks Unmasks Col. Giskes, Feb 1943

> Then, and only then, could a trap be set for the Germans which even SOE might regard as conclusive.
>
> — Leo Marks, *Between Silk and Cyanide* (1998), 159

Leo Marks was the son and designated future owner of Marks & Co, the world-famous second-hand bookseller at 84 Charing Cross Road where as a boy he'd broken his first code — the shop's business cipher. At age 22 Marks was a poet, a romantic, and a skeptic. He was a skilled crossword puzzler, contributing many to *The Times*. Most importantly, he was already a world-class cryptologist who'd independently reinvented in the theoretically unbreakable secure one-time pad (OPD) system of ciphering, which he called "worked-out keys" (WOKS).[63] And it was then, 1942, as SOE's chief cryptographer that

---

62 Marks (1998), 376.
63 Marks (1998), 33-35 and throughout. It appears (p.21) he'd made this breakthrough by Jul 1942.

he suddenly realized from irregularities in the clandestine radio traffic from SOE agents in German-occupied Holland that they'd been caught and were being controlled by German military intelligence (the Abwehr). That tragic case, Operation NORTH POLE, is summarized below (Chapter 4.8.2). Here, jumping ahead of that story, is the place to describe how Marks set a trap to force Abwehr Col. Hermann Giskes to reveal his dangerous game of double-agentry.[64]

Marks quickly became certain in his own mind that all these early SOE agents being sent into Holland had been captured and turned around by German Intelligence, i.e., Abwehr Col. Giskes. Marks based this conclusion on two facts. First, the frequent lack of "security checks" — more than could be reasonably attributed to the agents' haste or forgetfulness. Second, the many odd incongruities in messages and the manner of their transmission — as if the agents were hinting that they were sending under duress. Either of these facts was a yellow warning flag, the two together should have been seen as a red-flag hypothesis that demanded urgent verification. However, Marks couldn't get the first three successive chiefs of the SOE Dutch Section to accept his suspicions. In frustration, in early February 1943 Marks came up with what he privately called "Plan Giskes" to force the German puppet master to reveal himself:[65]

> I knew now how to give the Germans a chance to go wrong. The idea was dangerous and could easily backfire. I would have to wait for exactly the right moment before launching it. But at last I knew what the right moment was. ...
>
> [T]here was only one remedy, and it wasn't likely to arrive in time to stop the [next agent] team from being dispatched to Herr Giskes. No matter how costly the delay, I had to wait for the Dutch section to cancel a message to one of its agents. Then, and only then, could a trap be set for the Germans which even SOE might regard as conclusive.

The right moment came on 15 February when a routine reply message to agent Boni was cancelled by the SOE. Marks, entirely on his own, had a special indecipherable reply sent. It was designed to be so garbled that Boni would be beyond his skills to decipher. It would require a professional cryptographer. So if Boni replied, it would mean someone had deciphered it for him and that someone would have to be one of Giskes's staff. Boni replied. Giskes had sprung Marks' trap.

---

64 Marks (1998), index under "Plan Giskes" but specifically pages158-159, 203-208.
65 Marks (1998), 158-159.

Of course, as we'll see below in Case 4.8.2, the SOE Dutch section chiefs still resisted facing the fact that everything they'd been doing — the whole Dutch Resistance movement — had been controlled by the enemy. These chiefs had simply failed to follow the most basic rule of security to compartmentalize, to split into isolated "cells".

Ever since WW2, many Dutch, British, and other writers have proposed a conspiratorial theory of deliberate betrayal or "purposeful policy" by Britain — the deliberate sacrifice of these agents to protect some higher secret such as Normandy D-Day.[66] These writers overlook or discount the most likely hypothesis — that some British intelligencers has simply been grossly negligent.[67] See CASE 4.8.2 for a similarly devious conspiracy theory about SOE agents in France.

CASE 4.6.2:
The Dreyfus Ruse, 1899

The Dreyfus case of the 1890s is the most famous example of a loyal officer being framed for treasonous espionage. Captain Alfred Dreyfus was serving in the French Army in 1894 when he was accused of passing French military secrets to Germany & Italy. Convicted by a court-martial, he was sentenced to life solitary confinement on Devil's Island until 1899 when the French government finally admitted its error. His innocence was proved by showing the key cryptographic evidence against him to have been forged.

French cipher experts wished to clarify their ambiguous solution of an intercepted Italian encrypted message which seemed to indicate that Captain Dreyfus had betrayed French military secrets to the Italians. So the French had a double agent give the Italian military attaché a fake message that they'd carefully worded to enable them to recognize it in the attache's coded traffic to Rome. Because the attaché thoughtlessly encoded this message word-for-word, the French cryptanalyst's were able to verify their earlier decode, showing that the original message had been deliberately modified to frame Captain Dreyfus.[68]

---

66 The highly likely incompetence theory has been best argued by M. R. D. Foot, "The Dutch Affair," *Intelligence and National Security*, Vol.20, No.2 (Jun 2005), 341-343.

67 This much less likely lambs-to-slaughter theory has been best presented by Jo Wolters, "Remarks Concerning a Research Note on *The Dutch Affair*," *Intelligence and National Security*, Vol.21, No.3 (Jun 2006), 459-466.

68 Kahn (1967), 258-261.

CASE 4.6.3:
The Forensic Accountant vs the Embezzler, 1970s

> I have bit the biter.
> — Alexander Smith, *Memoirs ... of Jonathan Wild* (1726), 133

A decidedly non-normal forensic audit procedure was developed by one Canadian auditor. My friend, the late P. Howard Lyons, was a distinguished member of that profession, a former member of the Queen's Commission on Auditing in Canada and a senior partner in Deloitte, Haskins, & Sells. The fact that my friend was also a distinguished amateur magician as well as a bit of a practical joker probably enhanced his ability to detect fiscal fraud. Around the late 1970s he was called down to the States to investigate a case of suspected embezzlement in a major (Fortune 100) U.S. corporation. He set an unusually ingenious trap to catch the thief. First he prepared a fake edition of his firm's audit manual, including a false procedure for detecting error. Arriving at the American corporation, he left this special version of the manual on his VIP visitor desk. Days later, while going through the corporate books, he was amused to find that the bait had been swallowed. The books had been altered, incorporating Lyons' fake procedure. The bookkeeper, in his effort to conform to what be had been led to believe were the proper procedures to conceal his theft from the audit, had introduced a small but glaringly incongruent detail that not only confirmed deliberate embezzlement (rather than some mere slip-up in accounting) but pointed the finger of guilt straight at himself.[69]

CASE 4.6.4:
America's Midway Ruse, 1942

> The Battle of Midway, as Yamamoto saw it, would be the largest ambush in history.
> — Jeffery Richelson, *A Century of Spies* (1995), 180

> Midway was essentially a victory of intelligence. In attempting surprise, the Japanese were themselves surprised.
> — Admiral Chester Nimitz, *The Great Sea War* (1960), 245

---

69 Telephone interview with P. Howard Lyons, 26 March 1984.

U.S. Admiral Nimitz, Commander-in-Chief Pacific, suspected from codebreaking that Midway island was the most likely target of the next Japanese offensive; and he knew that offensive was imminent. Some recent revisionist naval historians claim that because of this he didn't need the provocative ruse described below. They overlook the ever-welcome and great value of fresh evidence that brings conclusive verification to a fragile hypothesis.

It was only six months after the U.S. Navy's catastrophe at Pearl Harbor. Japanese Admiral Yamamoto had proved to everyone including himself the incalculable value of surprise. He also knew that the American Pacific fleet was still incapable of simultaneously defending all of its vulnerable points. This gave Yamamoto a free choice of targets. Moreover, he knew that at whatever target he picked he would enjoy overwhelming strength. His pick was the small but strategic outpost of Midway Island. To his attack force he committed 11 battleships, 5 aircraft carriers, 16 cruisers, and 49 destroyers. Admiral Nimitz could muster an opposing force of only 3 carriers, 8 cruisers, and 14 destroyers.

Admiral Nimitz's only hope was to strike with a Pearl Harbor-level of almost total surprise. The story of how he did this is told by American amateur cryptologist David Kahn:[70]

> The question of where [the Japanese would attack] was answered fairly quickly by the Combat Intelligence Unit. The Japanese indicated geographic locations by maps with coordinates in code; they called these their CHI-HE systems, and they served as much to avoid [typo] error ... as to conceal. The cryptanalyst's had partly recovered one such map ....
> Several weeks earlier, they had discovered the code coordinates AF in a message sent from two [Japanese] scout planes over Midway. Context suggested that AF meant *Midway*. When they checked this against their partially solved map grid, they found that A's representing one coordinate of Midway's position and F's representing the other fit into it perfectly. So when they saw that AF was the codegroup for the locus of the main attack, they felt quite sure that Midway was the target.
>
> But the top brass squinted at this identification. On it rode the very existence of the American fleet and the future course of the whole Pacific war. They demanded confirmation.
>
> Rochefort [a crytographer] decided to trick the Japanese into giving him the proof. He cooked up the idea of having

---

70 Kahn (1967), 568-569. For fuller context see Jeffrey Richelson, *A Century of Spies* (New York: Oxford University Press, 1995), 182-183.

the Midway garrison broadcast a distinctive plain language message which would presumably be picked up by Japanese monitors.

Their coded report would be intercepted and solved by Americans, and the geographic indicator that they used in this telltale dispatch would have to mean *Midway*. Layton [Rochefort's colleague] liked the idea, and the two men drafted a message in which Midway reported that its fresh-water distillation plant had broken down. They cabled it to the atoll with an order to radio it back to Pearl in clear. Midway complied. The cryptanalyst's waited. Two days later there appeared in the harvest of Japanese intercepts one stating that AF was short of fresh water.

CASE 4.6.5:
Hypothetical Scenarios, 1984 to present

I'd written in 1977 a short story of an intelligence double-cross. It involved provoking the opponent's intelligence service into revealing which of two possible places they'd set an ambush.[71] This set me to wondering this past December what other types of intelligence might an opponent be induced to reveal by a counter-deceiver using a custom-tailored provocation. My specific hypothetical had provoked the truth about the PLACE (right or left flank) of an imminent ambush. What about the other categories in my 9-category checklist? These are: CHANNEL, PATTERN, PLAYERS, INTENTION, PAYOFF, PLACE, TIME, STRENGTH, and STYLE.[72]

I suspect that a suitable type of ruse could be devised to provoke information about each. One example must suffice:

How, for example, could one use provocation to reveal information about the PLAYERS category. Consider a poison-pen leaker in your organization. Specifically, a mole in the office who is leaking derogatory information about the boss to undermine his his position in the office political structure. Invent a few facts of the kind being used by the leaker — ones that would soon be contradicted and rendered harmless by developing facts. Then plant one of these fictions with each suspect and wait for the self-revealing comeback.

---

71 "The Door to Death: A Short Story of Counter-Deception." 1st draft, May 1977. Printed in Bart Whaley, *A Reader in Deception and Counterdeception* (Monterey, CA: 2003), 397-403; and reprinted in Whaley, Readings in Political-Military Counterdeception (FDDC, Dec 2007), Chapter 5.5.2 (pp.651-659).
72 As defined & discussed in Whaley, *Deception Verification in Depth & Across Topics* (FDDC, Nov 2009), 30. This list, which had begun in 1968 with the conventionally accepted TIME, PLACE, & STRENTH model, evolved through 1969 when it had 5 categories and 1980 with 8 until 1982 when it reached its present (and final?) 9 categories.

If you have a prime suspect in mind, you can save effort by inventing just one rumor that you reveal to her. Then wait for the comeback. If the rumor circulates back to you, then the leaker, the mole, the PLAYER is unmasked.

Incidentally, this example isn't hypothetical. I used it successfully in the mid-1980s to protect my then boss. To avoid causing further harm to him, I invented a derogatory story about myself — that I planned to take vacation at a crucial stage in the project. When, within a couple of days, that unique piece of information circulated back to me as office rumor, this identified who among the staff had leaked, who was the PLAYER. I leave it to the reader (or further systematic research) to imagine how provocation might apply to the other categories of deception.

4.7. Double-Bluffing

> "This is a trick of the kind so prettily described by conjurors as a 'Sucker'."
>
> — Charles Waller, *Magical Nights at the Theatre* (1980), 163

Magicians call this the Sucker Gag. It is one of their two most nearly undetectable types of deception (the other being the One Ahead Effect). A Sucker Gag or Sucker Effect or just plain Sucker is any trick where the magician leads the spectators to believe they have detected the method and then works a double bluff to surprise them even more.

CASE 4.7.1:
The Booby Trapped Booby Trap: Major Foot's Luck, 1943

One of the cliché suspense dilemmas in short stories, movies, & TV is of the bomb disposal expert who suspects that the wiring on the device in hand may a double-bluff — designed not to defuse but to trigger. Unfortunately, this is equally a real-world dilemma for all those who seek to disarm land mines and booby traps. Specialists call them "anti-handling devices".[73]

The earliest systematic large-scale use of these traps within traps is attributed to the German Luftwaffe with its ZUS-40 anti-removal bomb fuze introduced during the London Blitz in 1940. The later, most diabolical German models, cunningly added a *second* anti-handling fuse that looked differently and worked independently of the first.

---

73   *Wikipedia*, "Anti-handling device" (accessed 14 Dec 2009).

When in 1999 I met Professor M. R. D. Foot, I didn't know I owed that opportunity to his luck when in WW2 he encountered one of those German devices. In early July 1943 Major Foot was serving as the British SAS Brigade intelligence officer with a series of eight Commando raids (Operation FORFAR) on the German-occupied French coast. He'd been told that the mission of the raid he went on was to capture German soldiers. He hadn't been told that FORFAR's real mission was a small part of a major (but unsuccessful) deception operation (STARKY). That operation had been intended to suggest an imminent but fictitious Allied invasion of northern France as cover for the real invasion of far off Sicily.[74]

This one raid came up empty. Their target, a German pillbox on the coast, was unmanned at that time. All they brought back was a new German anti-tank mine. Major Foot took this bit of booty to M110, the department charged with studying enemy equipment. Then, as he recalled:[75]

> They took one look at the mine and the senior officer present bellowed "Clear the room!" I said, very casually, "Look, it's made safe. There's a nail through the fuse exactly as you recommended." He yelled again, "Clear the room!" So we cleared the room and one brave boy went back in. He came back out a few minutes later and said, "It's all right now, sir."
>
> I discovered that it was a new type of terror mine which I hadn't met before. If you turned it upside down, it went off. Completely by chance I had carried it vertically and I hadn't turned it upside down, so it hadn't gone off. Just as well, really.

CASE 4.7.2:
The CORONA Satellite That Wasn't, 1958

Amrom Katz and Merton Davies dreamed of a world-wide satellite reconnaissance system. That was in the late 1950s when both men were with the RAND Corporation as experts on aerial photo-imagery. The struggle for dominance of outer space became urgent in October 1957 when the Soviets surprised American intelligence by successfully launching Sputnik-1, the first space satellite.

The conventional design would have kept the photo-interpreters waiting for days or weeks until the entire satellite would return to Earth with its film payload intact. Katz & Davies came up with a workable system whereby the exposed camera film would be ejected in orbit in a parachuted-fitted pod. This would then be scooped up mid-air by a recovery plane. This was intended as a

---

74 On FORFAR see Holt (2004), 487, 820; Barbier (2007),113.
75 Foot in Young & Stamp (1989), 76.

stop-gap measure during the year or so it was assumed it would take to develop a near-real-time radio-imagery system.[76]

Katz and Davies represented RAND at the February 1958 monthly review meeting between the USAF and its contractors on the reconnaissance project. When Lockheed's project manager announced without explanation that the film recovery program had been cancelled Katz and Davies leapt to their feet and, as one participant recalled:[77]

> They went ballistic. Amrom particularly took it upon himself to try to get the effort reinstated and he began going around the country briefing anyone who would listen about the unwise decision to cancel the recoverable camera system. I mean he had a cause! He became so well known as an agitator on this that he disqualified himself for being cleared for what was now a black program — even though he had conceived of it! The folks in charge knew that if they cleared Amrom, he would immediately cease agitating and that would tip everone else that the program was [still] underway.

The Katz-Davies concept of a film eject & recovery system worked. Run beginning 10 Mar 1958 as a "black" program under the CIA's Richard Bissell and renamed CORONA, it had its first successful flight on 28 Feb 1959. Fortunately, because CORONA proved to be more than a stop-gap measure of a few months. In fact, it would be more than a decade before the alternate system — near real-time radio data transmissions — would be provided by the follow-on KH-11 "Keyhole" satellites.

The CORONA ruse apparently succeeded in deceiving the Russians. However, the deception of Katz & Davies was another matter. Knowing that Amrom would have relished playing a game of double-bluff, I doubt it had been wise to keep him unwitting, sidelined, & unhelpful and suspect the decision to do so was more a matter of petty dislike than pragmatic security.

CASE 4.7.3:
Jones and the Fake Jay Beam, 1941

Dr. R. V. Jones was already a world-class practical joker when early in WW2 he became the British Air Ministry's chief of Scientific Intelligence. Thus he arrived in his new job fully prepared to plan the most sophisticated types of deception. He could skip the prolonged phase of trial-and-error on-the-job-learning that marked almost all other WW2 deceptionists. Indeed, the only

---

76 Curtis Peebles, *The CORONA Project: America's First Spy Satellites* (Annapolis: Naval Institute Press, 1997), 27-30, 42, 45.
77 Harold F. Weinberg interview with Hall, 16 Mar 1995, as quoted in Hall (1998), 114. See also Peebles (1997), 45.

notable exception was Col. Dudley Clarke who as an amateur magician, was similarly "off and running" in his wartime assignment as the British Army's chief of deception in the Middle East.

Consequently, Jones's first affort at military deception was about to be that highly sophisticated type known as a Double-Bluff (Sucker Gag). This case occurred at the start of the lengthy cat-&-mouse game throughout the European phase of WW2 between Dr. Jones and General Wolfgang Martini as the Luftwaffe's Director of Air Signals. This was the game of successive electronic countermeasures (ECM) and counter-countermeasures (ECCM) that came to be known as the Battle of the Beams. Jones's opening salvo in this battle was the case of the Gee-Jay Beam.

The Gee Beam was the RAF's latest radar navigational device. When in August 1941 a crashed RAF bomber fitted with an experimental model of Gee crashed — and presumably recovered by the Germans — Jones decided to use flattery to mislead them — by persuading the Germans to think the RAF had copied a German invention. Accordingly, to mask Gee's real function, Jones planted clues implying that Jay worked along different principles from Gee. Then, to make them believe some other secret device was giving the navigational guidance, he planted a different set of clues. And, finally, he camouflaged Gee's real transmitting stations in England to look like standard radar stations. This hoax worked for nearly 6 months before Germans figured it out.[78]

CASE 4.7.4:
Jones and the Double-Telltale Envelope, early 1940s[79]

Dr. R. V. Jones gives an example of a telltale with an instructive twist. During WW2 a colleague suggested a simple way to detect whether their own confidential letters had been opened without authorization. He explains: After sealing an envelope in the ordinary way, stitch together the envelope and its enclosed sheets in a sewing machine. This will make match-up of the holes and reassembly with new thread quite difficult for any "seals & flaps" specialist. Jones added his recommendation that the common cotton sewing thread be marked at intervals with fluorescent dye. "I banked," he explains, "on a human weakness that I had seen several times before (and many since) which beguiles clever minds to concentrate on the difficult parts of a challenge to such an extent that they overlook the simpler ones." Three days later the test letter was returned by the British security experts, all properly stitched and therefore seemingly unopened. But the security lads had enclosed their visiting cards

---

78   Jones, *Reflections on Intelligence* (1989), 135-140; Jones, Wizard War (1978), 217-222; Moss (1977), 22-23.

79   As originally in Whaley, *The Maverick Detective* (manuscript) and then in Whaley, *Textbook of Political-Military Counterdeception* (FDDC, Aug 2007), 92.

to prove they'd opened it. They were "furious", Jones recalls, when told they'd overlooked the telltale fluorescent thread.[80]

CASE 4.7.5:
The Allies Double-Bluff the German Gothic Line in Italy, 1944

> The first and only full-scale double bluff ever mounted by "A" Force.
>
> — Holt, *The Deceivers* (2004), 620

The Allied final drive up the Italian peninsula had been stopped in the mountains just short of the open plains of the Po Valley during the summer of 1944 by a combination of the formidable terrain and fierce German resistance. The Allied commander, British General Sir Harold Alexander was set to unleash the breakthrough Operation OLIVE offensive. His A-Force deception planning team had done their work.[81]

For his decisive push to breech the German-held Gothic Line (Gotenstellung)[82] that reinforced the natural defenses of the Apennines, General Alexander had originally (in early June 1944) planned his main breakthrough to go through the mountainous center, with a feint along the Adriatic coast. The deception plan (Operation OTTRINGTON) — worked mainly by radio traffic & 100+ dummy tanks plus support vehicles — had convinced his German counterpart, Field-Marshal Kesselring, that the Adriatic feint would be the main attack.[83] Then, on 5 July, Alexander was formally notified that the final decisions in Washington and London to proceed with the upcoming landings in southern France would require a change in his plans. Alexander would lose most of the units he needed for his offensive. Not just many of his American divisions; but, particularly, all his French divisions on whose mountain warfare experience he'd been counting for the breakthrough.

Consequently, on 4 August, Lieutenant-General Oliver Leese, the British Eighth Army commander, convinced Alexander to reverse his original plan, making the real attack up the Adriatic coast with Eighth Army while the denuded Fifth Army conducted the demonstration attack at the center, above

---

80   Jones (1989), 37.
81   This case has been condensed from Whaley, *Stratagem* (1969), CASE B38 (Battle of the Gothic Line, 25 Aug 1944: Operation OLIVE); and supplemented by Holt (2004), 620-621. See also Douglas Orgill, *The Gothic Line* (New York: Norton, 1967); W. G. F. Jackson, *The Battle for Italy* (New York: Harper & Row, 1967), 252-277.
82   Although the Germans had already renamed it the "Green Line," the earlier designation is more familiar.
83   Orgill (1967), 30. But compare Albert Kesselring, *A Soldier's Story* (New York: Morrow, 1954), 256.

Florence. In other words, *adopt the very strategy that the current deception operations were successfully communicating to the Germans.*[84]

This reverse of strategy now required a plausibly adjusted stratagem. Moreover, with OLIVE D-day set for 25 August, Alexander's deception planners had only three weeks to reverse the perceptions of the enemy — a feat that history shows to be extremely rare. And one that I believe to be one of the most difficult type of deceptions to carry out.

The essence of the new deception plan (Operation ULSTER) — designed at Caserta by A-Force's Col. Dudley Clarke — was to work a double-bluff. This made the Germans now believe that the old "evidence" (OTTRINGTON) had been and still was part of a deliberate bluff — the essence of the gambling cheat's Shell Game or Three-Card Monte. This was attempted by a new and elaborate program of radio deception. It consisted, at least in part, of using radio to simulate the continued presence in the center of the departed divisions of V Corps).[85]

As far as I am aware, this is the only case where a military deception plan was turned back upon itself — a stratagem to attempt to discredit earlier disinformation by exposing it for the stratagem it was.

The shift in the axis of attack from center to east involved transferring two Eighth Army corps totaling eight divisions from the center to the east coast behind the two divisions already manning that part of the line. This transfer was completed by D-minus-5 without the Germans quite realizing what had happened.[86] Headquarters of the opposing German Tenth Army had received reports of the noise and dust created by the large-scale Allied lateral shift. Army headquarters was interested but tended to accept the opinion of the commander of its 76th Panzer Corps, General Traugott Herr, that this movement was merely the *westward* shipment of war material that he assumed was now being unloaded in quantity at the recently captured Adriatic port of Ancona.[87] Allied air supremacy prevented the Germans from verifying Herr's false hypothesis.

In any case, Kesselring was not entirely disabused about the *original* deception plan and had, accordingly, reinforced Tenth Army with General Richard Heidrich's crack 1st Parachute Division. Nevertheless, Kesselring did perceive a center attack as sufficiently plausible that he insured himself against that alternative by holding two divisions in reserve near Bologna.[88]

---

84 Jackson (1967), 266.
85 Orgill (1967), 33.
86 A vivid eyewitness report by *Collier's* war correspondent is Martha Gellhorn, *The Face of War* (New York: Simon and Schuster, 1959), 116-175.
87 Orgill (1967), 33-34.
88 Orgill (1967), 33.

When OLIVE began at one hour before midnight — with five divisions jumping the Metauro River and spearheading the three-corps attack — the usual preceding artillery preparation was dispensed with in order to insure maximum surprise. And surprise was indeed achieved, not only of place and strength but of timing as well — both the German field commander (Colonel-General Heinrich von Vietinghoff) and a key divisional commander (General Richard Heidrich) were still on leave.[89] Moreover, the assault caught one German division just during the period it was being relieved in the line.[90]

The ULSTER double bluff had only been a partial success, in part because German Intelligence had intercepted a copy of an Eighth Army commander's genuine message to his troops. Despite ULSTER's marginal success, Col. Clarke remained leery of this particular type of double-bluff deception where one real deception plan is suddenly switched for another. That's just too tricky, too risky, too uncertain.[91]

Surprise gave initial success on the ground, but the offensive was soon halted by the combination of the difficult terrain, lack of reserve units, and the ability of the Germans to bring theirs up over a far better road system.

CASE 4.7.6:
RFE Turnabouts a Czech Terror Plot, 1957

> "The Press, Watson, is a most valuable institution, if you only know how to use it."
>
> — Sherlock Holmes in "The Adventure of the Six Napoleons," *Collier's Weekly* (30 Apr 1904)

A marginal case of turnabout was the attempt by the Czech Intelligence Service (StB) in December 1957 to demoralize the employees at Radio Free Europe in Munich by planting a saltshaker filled with poison-laced salt in the radio station's cafeteria. The intent was to terrorize the employees, most of whom were East European refugees, by making a few violently sick enough to make all of them realize how vulnerable they were to real assassination. However, the Czech RFE employee selected by StB officer Josef Frolik to plant the shaker was a CIA controlled agent. So the plot failed.

But the game wasn't over. The CIA/RFE officers involved were clever enough propagandists to take advantage of the Czech's game. They now played

---
89 Orgill (1967), 43.
90 Kesselring (1954), 256. Compare Orgill (1967), 43, who in citing British sources says two German divisions were being relieved.
91 Holt (2004), 620, 1044n620.

turnabout by trumpeting it to the world's press as a Communist terrorist mass assassination plot foiled by tight RFE security. Consequently it was the Czechs who were publically embarrassed worldwide.[92]

That's the full version of this odd incident. Interestingly, RFE's cleverly spun version is still the one that dominates the many Internet references.

4.8. Double Agentry

> The [British deception] planners had the powerful channel of deception provided by MI5's success in rounding up virtually all the German agents in Britain, and in "turning" some of these to send false or misleading information.
>
> — R. V. Jones, *Reflections on Intelligence* (1989), 124

Double-cross agents are a special channel of information, one designed to feed disinformation directly into the opposition's intelligence system. Double agents have been a staple of the counterintelligence & counterespionage scene since antiquity, having been mentioned as early as around 350 BC by Sun Tzu. His famous *Principles of War* identified the double among his five different types of secret espionage agent:

- Native [recruited locally]
- Inside [moles, penetration agents]
- Turned [doubled, double agents]
- Dead or Expendable [primed with false information & then sent to be captured]
- Living [spies sent to scout & return with information]

Taken together, as Sun Tzu clearly intended with his imagery that they form a "Divine Skein", like a fishing net that lands all the fish with a single pulling together of its cords, these five are the "ruler's treasure" — what a perfect definition of HUMINT. Here, of course, our interest is focused on Master Sun's early appreciation of the double agent as an example of double-cross.

Evidently all pre-WW2 cases of double agentry employed doubles on a one-by-one, case-by-case basis. The first coordinated interlocking teams of double

---

92 Ladislav Bittman, *The Deception Game* (Syracuse: Syracuse University Press. 1972), 11-12; BW interview with Bittman, 1971; and Theodore Shackley, *Spymaster* (Dulles: Potomac Books, 2005), 21-22.

agents was apparently the Double-Cross System developed during WW2 by the British.[93] Incidentally it had been invented as an orchestrated system of agents by Col. Dudley Clarke as a key part of his A Force deception team in North Africa & the Near East in 1941 and only later adopted in the more famous work by Masterman and his Double-Cross Committee in London.[94] These two British double-agent systems are a fine model of deception run at considerable depth, great breadth, and with tight coordination.

Further details and comment are unnecessary here, as all intelligence analysts — even those not particularly keyed in on deception — are surely familiar with this deservedly famous counterespionage operation.

During the middle and end game of WW2, the British became — through many trials and few errors — the world's best at running a double-agent system. Their's grew into a vast system, a veritable symphony orchestra of nearly 200 players who dominated the main venues from London, across Europe and the Middle East, to New Delhi. Because of its sheer size, tight coordination, and complete dependence for both its success and fine-tuning on the feedback that came from superb reading of the German intelligence codes, we are told that it was unique and would remain so. That may be true. But no one seems to have examined the probabilities of constructing a less ambitious system. For further discussion on this point see Chapter 7.

The story told here shows its slim beginning in North Africa as a template in the mind of Colonel Dudley Clarke. It has been selected because it shows that this is a deception and counterdeception game that can be developed and worked even on a small scale and by a subordinate unit.

CASE 4.8.1:
CHEESE: World's First Double-Agent System, 1939-1944

Double agents have been commonplace tools of intelligence services since at least as early as the time of Sun Tzu in ancient China. But these were individuals, operated independently of any other double agents. The first double-agent system, where several double agents are coordinated as part of a single deception operation or campaign, was conceived and operated in the Near East in WW2 by Col. Dudley Clarke. Begun in 1939 with a single double agent dubbed CHEESE, Clarke developed it into a full-fledged system.[95] The

---

93 Marginal exceptions might include the Soviet-run Trust in the 1920s or their post-WW2 Albanian & Ukrainian traps as well as the British thorough infiltration of the IRA Provos in the 1990s. Even the FBI's post-WW2 penetration of the CPUSA might qualify except that Director Hoover ran it only for collecting *intelligence* and not as a deception (double-cross) system.
94 Mure (1980), 65-79; Masterman (1972); Hesketh (1999), 46-57; Harris (2000); Crowdy (2008).
95 Mure (1977), index under "Cheese" & 254; Mure (1980), 65-71, 270; Delmer (1971), 32-35, 40, where he is called "Orlando"; Crowdy (2009), 137-139, 169, 184-185; Rankin (2008), 333-334, 345; Holt (2004), index; Masterman (1972), 107.

following more familiar cases are just minor variations on Col. Clarke's brilliant theme.

CASE 4.8.2:
NORTH POLE: German Playback Double-Agentry, 1942-1944

> I had a gut feeling about the Dutch but it was a bride unwilling to be carried over the threshold of consciousness, and I couldn't pinpoint it. ... There was something wrong with the Dutch traffic.
>
> — Leo Marks, *Between Silk and Cyanide* (1999), 98, 102

> One single control group, dropped blind and unknown to us in Holland, could have punctured in an instant the whole gigantic bubble of Operation Nordpol.
>
> — Hermann Giskes, *London Calling North Pole* (1953), 100

Operation NORTH POLE (Unternehmen NORDPOL aka ENGLANDSPIEL) was a large-scale and sophisticated counterespionage "radio game" played against the British Special Operations Executive (SOE) by Abwehr Major (later Lieutenant-Colonel) Hermann Josef Giskes. Giskes captured his first Dutch resistance agents operating in the Netherlands and induced them into working for him as double agents. These first agents then arranged for SOE to airdrop additional agents, radios, and supplies — all of which were captured on landing and most added to Giskes's growing network of double-agents. Beginning on 6 March 1942 (with the capture of Lauwers) Giskes ran this game for 20 months until October 1943. Then the British cut off supplies (largely because the RAF had become tired of losing so many planes & crews). Thence the radio game show dragged on until 1 April 1944 when Giskes signed off on April Fool's Day with a sarcastic taunt.

This system fooled the Allies into continuing the airdrops, which comprised 190 drops of new agents plus providing the earlier ones with information and supplies, including large enough quantities of munitions and money to equip the entire Dutch Resistance. In all, 54 SOE Dutch agents were captured & run as double agents and 47 eventually executed. Additionally, 12 RAF heavy bombers had been shot down and 40 aircrew killed during resupply missions. For over a year Major Giskes was, in effect, the Commander-in-Chief of the Dutch Underground.[96] Former CIA Director Dulles would rate this "one of the most effective German counterespionage operations of all time."[97]

How could this have happened? Captured radio operators continued broadcasting encrypted messages, but without the required "Security Checks" that had been designed to indicate they had been captured and were transmitting under German control.

Further, SOE's astute code chief, Leo Marks, soon noticed that, contrary to SOP, the Dutch messages contained none of the common errors that would have made them indecipherable. Marks correctly inferred this was because they had not been coded hastily in the field by nervous agents, but by unhurried German cryptographers. To test this hypothesis, Marks sent a deliberate indecipherable. When it was answered, he again correctly inferred that no SOE agent had the skill needed to have reconstructed his message. Nevertheless, Marks's superiors took no action.

Moreover, when in 1943 two Dutch agents managed to escape and return to Britain, their story was disbelieved and they were arrested for suspected counter-espionage. To discredit them by "proving" the Dutch Resistance was free of control, Giskes had organized what appeared to be a major work of sabotage in Rotterdam harbor, the spectacular noontime blowing up of a large vessel — actually just an old barge loaded with damaged aircraft parts.[98]

In December 1943, Major Giskes correctly inferred that SOE had finally figured his game and were trying to deceive him. Having picked April 1st as a suitable day for a practical joke, he sent a facetious message over all 10 of his controlled radio links. Thus ended this great deceptive counterespionage radio game.

Other than the usual unwillingness to accept awkward facts, a major reason the SOE chiefs had been so slow to realize all their Dutch agents were under complete enemy control was inter-departmental rivalry — between individual country sections, between the country sections and the central code section, and most importantly between the SOE and the other British intelligence services.[99]

---

96 Giskes (1953); Marks (1998), index; David Kahn, *The Codebreakers* (Revised, New York: Scribner, 1997), 531-538; Philippe Ganier-Raymond, *The Tangled Web* (London: Barker, 1968); M. R. D. Foot, *Holland at War against Hitler* (London: Cass, 1990). A fairly accurate summary is *Wikipedia*, "Englandspiel" (accessed 23 Dec 2009).
97 Allen Dulles, *Great True Spy Stories* (New York: Harper & Row, 1968), 364.
98 Giskes (1953), 112-114; Marks (1998), 431.
99 One of the more insidious false rumors of WW2 that persists to the present is that the British had deliberately betrayed their Dutch agents to the enemy for deception purposes. This rumor has both its origins & persistence in the widely held misconception that the so-called British Secret Intelligence Service (SIS or MI6) was omniscient & competent. For those deluded

A similarly false theory at that time was that SOE's problems in France were caused by a traitor in London, as well-disproved by Foot (2004), 115.

CASE 4.8.3:
LAGARTO: Japanese Double-Agentry, 1943-45

The Japanese worked a similar double-agent "radio game" against the Allies in the Pacific. For 22 months, from Oct 1943 to Aug 1945, Japanese Intelligence played back to MacArthur's Australia-based inter-allied Services Reconnaissance Department (SRD) an Australian Army radio agent. This successfully compromised a small-scale Allied covert operation in Japanese-occupied Portuguese East Timor. Operation LAGARTO was, on a smaller scale, a virtual replay — including the final insulting "up-yours" sign-off — of Operation NORTH POLE.[100]

CASE 4.8.4:
SCHERHORN: A Soviet Playback Operation, 1944-1945

In July 1944, on Stalin's personal order, a small Soviet intelligence unit on the Eastern Front began Operation SCHERHORN by transmitting a request to the German High Command on behalf of Lt.-Colonel Heinrich Scherhorn. The deception story had the German elite believing that Scherhorn was surrounded and cut off in Byelorussia with his 2,500-man armored brigade. This message and later ones brought a succession of parachuted medical supplies, small arms, food, money, and 25 German agents & Polish guides. The German High Command had confirmed that Scherhorn was, indeed, one of their unit commanders who'd gone missing in action. Attempts to verify the information were difficult. What had been confirmed, though, seemed to substantiate the Scherhorn story. Not only were the supplies & Polish guides provided by air drop to eagerly awaiting Russians, but Scherhorn was considered a hero and promoted to full colonel. Meanwhile, he was in fact a POW collaborating with the Soviet NKVD.[101]

This elaborate "radio game" was played successfully by Soviet intelligence until the very end of the war. The deception aspect came to light only when Colonel Scherhorn (who'd been promoted by Hitler on 23 March 1945) was eventually repatriated.

---

souls it's understandable that they found it easier to attribute the NORTH POLE tragedy to "perfidious Albion" than to simple incompetence. This rumor is adequately refuted by Marks (1998); M. R. D. Foot, *Holland at War against Hitler* (London: Cass, 1990); Lauwers in Giskes (1953), 200.

100 Anonymous, "A Small South Pole," *Studies in Intelligence*, Vol.4, No.4 (Fall 1960), A23-A27.

101 Dieter Sevin, "Operation SCHERHORN," *Military Review*, Vol. 46, No. 3 (Mar 1966), 35-43; Pavel & Anatoli Sudoplatov, *Special Tasks* (Boston: Little, Brown, 1994), 167-170.

CASE 4.8.5:
X-2: The American Way of Double-Agentry

During WW2 the American OSS's semi-equivalent of the British MI-5 Double-Cross (or XX) Committee was called X-2. I say "semi-equivalent" because it was more conventional counterintelligence than specialized in double-agent games. Nor did the American ever operate double-agents as a coordinated strategic system as had the British. Still, the experience was more than a footnote in the history of American intelligence.[102]

4.9. Tripwires & Telltales

> There were dozens of ways of planting telltales. A hair lightly stuck across the crack of the door was the most simple; a scrap of paper jammed against the back of a drawer would fall out when the drawer was opened; a lump of sugar under a thick carpet would be silently crushed by a footstep; a penny behind the lining of a suitcase lid would slide from front to back if the case were opened.
>
> — Ken Follett, *Triple* (1979), Ch.8

It's sometimes possible to trick the deceiver into self-betrayal. We do this by actively intervening, changing our deceiver's view of reality in order to provoke them into revealing new, key evidence that will disclose the deception or parts of it. We do this by means of invisible telltales or tripwires and psychologically baited traps.

Tripwires & telltales serve one of three purposes. First, benign, as with the door to a shop that triggers a bell or buzzer when opened to alert the occupant that a customer has entered and may need assistance. Second, defensive, such as a loud klaxon or flashing lights that warn of an unwelcome intruder. At best it drives them off; at worst it buys time to take defensive action. Neither type is relevant here. The third type, the silent alarm, is relevant.

Silent alarms enable us to retake the initiative, seize the offensive, and thereby surprise the intended surpriser. These silent alarms include both literal tripwires and figurative ones, such as motion detectors, computer intrusion

---

[102] Major sources on X-2 are Naftali (2002); Winks (1987), index; Richard Cutler, *Counterspy: Memoirs of a Counterintelligence Officer in World War II and the Cold War* (Washington, DC: Brassey's, 2004). A revealing novel about the running of double agents by an X-2 veteran officer in France is Edward Weismiller, *The Sleeping Serpent* (New York: Putnam's, 1962).

detectors, and invisible surveillance cameras. Only two conditions need be met: 1) The stealthy intruder is unaware of having triggered the alarm; and 2) the one intruded upon springs a trap or ambush. This sets the stage for a whole new set of rules — the defender's rules.

A nice case example from ancient Hebrew literature will suffice, one where the silent telltale was dust:

CASE 4.9.1:
Daniel & the Priests of Bel, c.100 BC[103]

In this legend our deception analyst is personally involved in a power struggle at the highest level of politics and uses a telltale to trick the tricksters. This story, most likely written shortly before 100 BC, was added late to the *Book of Daniel*, which, although accepted in the Vulgate version of the Old Testament, is now widely judged apocryphal.

The setting is Babylon soon after 539 BC when it was conquered by Persian King Cyrus the Great. There Cyrus has been converted by observing seemingly miraculous events in the Temple of Bel (or Baal), presided over by the huge clay and metal statue of that city's mighty god.

Most convincing to Cyrus was Bel's Miracle of the Feasts. Costly donations of fine food & drink are placed upon Bel's great altar table at night. Then priests and the faithful leave, closing the temple doors behind them. Next morning, when the temple reopens, the food is seen to have vanished.

Jewish prophet Daniel, suspecting priestly hocus-pocus, is unimpressed and refuses to worship Bel. King Cyrus seeks to convince the delinquent Jew by pointing out the vast quantities of food supernaturally consumed every night by Bel; but Daniel asks for a test — what today we'd call a controlled experiment.

Cyrus agrees to Daniel's test conditions but sets a forfeit — death for failure: for Daniel if he cannot prove deception and for the 70 priests if he proves them to have deceived their king. Both Daniel and the priests accept this mortal challenge, each confident of success.

That night, closely and openly watched by Cyrus and Daniel, the priests heap the food and wine upon the sacred altar table set before the great image of Bel and then leave. Daniel moves through the vast stone hall, apparently satisfying himself that it is indeed empty. Cyrus and Daniel step outside, and the single entrance is closed and the double-doors sealed, marked with the crests of both

---

103 As originally in Whaley, *The Maverick Detective* (manuscript) and then in Whaley, *Textbook of Political-Military Counterdeception* (FDDC, Aug 2007), 93.

king and priests. The several parties retire for the night, each with their own thoughts of the morrow.

At dawn, Cyrus and Daniel return to the temple. The great doors are still closed with seals intact. Yet, when the doors are thrown open, all standing outside can see that the altar table has been stripped clean. Cyrus declares that the priests have proven the miraculous powers of Bel. But Daniel invites the king to "Come, see the deceptions of the priests." They enter the empty hall, and Daniel points to the floor. There, with the morning sunlight streaming across it, Cyrus discerns the faint but numerous crisscrossing of bare footprints. On closer inspection, Cyrus sees that these cluster around the altar table and lead to and from a small area on the floor beneath the table, which is found to be a trap door. During the night, the 70 priests with their wives and children had secretly reentered the temple to consume the feast of Baal and leave their tell-tale footprints in the fine wood-ash that Daniel had sprinkled over the floor during his earlier "inspection". Outraged, King Cyrus orders the priests, wives, and children slain and gives the temple of Bel to Daniel that he may destroy it.[104]

CASE 4.9.2:
Noor's Secret Triple Security Checks, 1943

> She'd let her back be broken rather than tell a lie.
>
> — Leo Marks, *Between Silk and Cyanide* (1998), 320, of the doomed SOE secret agent, Noor Inayat Khan.

There are three common telltales that a message sent from a hostile place by one of our covert agents is authentic — that is, a) from our agent and b) not composed under duress or composed & transmitted by some impostor:

- First, in the old days of telegraphic (wire or radio) transmission of dot-dash Morse-type code, each operator's tempo & touch ("fist") on the telegraph key was as recognizable to other telegraphers as their voice. Intelligence services routinely had their Morse operators record test messages during training to give HQ a baseline reference for verification. Although this forensic technique was optimistically

---

[104] Detective fiction fans will recognize this as the second oldest "Locked Room Mystery," a popular sub-category in the mystery genre in modern times. The first, which dates from the 5th century BC is Herodotus's tale of "Pharaoh Rhampsinitus and the Thief," which also includes a fine example of deceiving the deceiver. An admirably accurate & comprehensive survey of the LRM is in *Wikipedia*, "Locked room mystery" (accessed 20 Jan 2010).

called "fingerprinting", captured agents "fists" were often convincingly imitated by skilled enemy imposters.[105]

- Second, the style in which the message was composed — any out-of-character words, spelling, grammar, phrasing, etc. Any major departure suggests we are dealing either with an impostor or a very brave agent trying to warn us that she's been caught. More subtle deviations are best handled by specialists in stylometry.[106]

- Third, and this particular telltale is the subject of this case It was standard procedure that each agent sent into enemy territory have a "security check" — a prearranged code meaning "I'm still free." By omitting this check the operator or an impostor is saying "I'm a prisoner." This security check system, of course, soon became known to opponents, so the idea of the "bluff check" was added.[107]

Noor Inayat Khan was the first and perhaps still unique exception to the matter of these security checks. Born in Moscow to a leading Indian Sufi mystic father and an American Buddhist mother, she was raised in England & France, where she studied child psychology & music at the Sorbonne. She'd become a skilled musician & published writer of children's stories by June 1940 when she fled from the Nazi invaders to England. Six months later she was assigned to the RAF and trained as a WAAF radio operator — a skill she quickly mastered.

After volunteering to be sent into Nazi-occupied France to work with the French Resistance underground she was accepted into the British Special Operations Executive (SOE) in Feb 1943. SOE trained her in clandestine tradecraft — at which she was both remarkably clumsy & forgetful — but was kept on because of the urgent need for an agent wireless operator fluent in French.[108]

Then came her training in the codes needed for messaging SOE HQ in London. Her trainer was SOE's chief cryptologist, Leo Marks. He was disturbed, challenged, and uneasy by this otherworldly and gentle creature — the least likely candidate for a secret agent being prepared to operate behind the lines of a clever enemy. His concern was doubled when he found that she was a

---

105 Foot (2004), 101-102; Marks (1998), 601-602.
106 Erica Klarreich, "Bookish Math: Statistical tests are unraveling knotty literary mysteries — stylometry," *Science News*, Vol.164, No.25/26 (20/27 Dec 2003), 392-394. Gives a clear explanation & summary of the stylometric method for attributing authorship. Shows stylometry's great strengths, particular weaknesses, and practical limitations.
107 Foot (2004), 99-101; Marks (1998), 16, 319-320; Lauwers in Giskes (1953), 184.
108 I don't recommend *Wikipedia*, "Noor Inayat Khan" (accessed 6 Jan 2010). The most relevant biography for our purpose is Foot (2004), 297-301. Although elements in her story from capture to death have become romanticized legends by well-meaning biographers, untrustworthy witnesses, and self-interested bureaucrats, most parts have solid evidentiary credentials, particularly as documented by Foot and recalled by Marks.

compulsive truthteller, one who would truthfully answer all questions, even those from strangers with no power over her.[109]

Her life-threatening ethics reached their limit the day Marks explained her two "security checks" — the "bluff check" and the "real check":[110]

> Like all agents using the poem-code, she had a "bluff" check which she was allowed to disclose to the enemy, and a "true" check which was supposed to be known only to London. The only circumstance in which the checks had any value was if the agent were caught *before* passing any messages, as the enemy had no back traffic to refer to.

When Marks explained that for these checks to work, she'd have to lie about them to her captors. She balked. Marks, flustered, had a sudden inspiration, one that bypassed her ethical bind. On the spot, he invented and gave her a completely new kind of security check, explaining:[111]

> All you have to do is remember one thing. *Never* use a keyphrase with eighteen letters in it — any other number but not eighteen. If you use eighteen, I'll know you've been caught.

She agreed and was rigorously trained by him to follow this protocol in a long testing session.

Marks, unfortunately, doesn't explain why this ruse works — perhaps he thinks it's either obvious or irrelevant. As it's the crucial lesson of this case, I'll reconstruct it:

The German interrogators had become conditioned to ask about the bluff+true checks. And they'd probably expect the next evolutionary step to be a bluff+bluff+true check. Now, this logic model requires that the agent-encoder make the *positive act of including* all bluff & true checks, however many of them might be in that agent's protocol.

So Marks designed Noor's personal check to work by a different — indeed opposite — logic. Her special "distress" check required always taking the *negative act of omitting* any 18-letter key. The enemy decoders would be unable to detect this incongruity in her first message sent under their control — they'd need at least two to even form an initial hypothesis about what she was doing. It was, therefore, unlikely her interrogators would know the right questions to ask. Indeed, they wouldn't even become aware that they'd slipped up — at least not until after that first message had already betrayed their presence to London.

---

109 Marks (1998), 317-321.
110 Marks (1998), 319.
111 Marks (1998), 320.

This procedure fits the general Principle of Incongruity that, when applied to impostors, I define as the following:[112]

> **RULE: Impostors never *conclusively* reveal themselves by things they don't know or fail to do — only by stating one thing they shouldn't know or by doing one thing they shouldn't be capable of.**

Note that this protocol is an entirely different means of ID than either passwords (face-to-face or as access codes) or the aviator's IFF (Identification Friend or Foe) radio signals. Both are of the "vocabulary" or element type rather than Marks's "grammatical" or structural.

■ ■ ■ ■ ■

Noor, now code named "Madeleine" and traveling in a small RAF Lysander liaison plane, landed in France on the night of 16/17 Jun 1943 and began working in Paris with the Prosper circuit (cell). During the next few weeks German counter-espionage managed a major disruption of this key Resistance group by capturing all the other Prosper radio operators. As Marks recalled:[113]

> Knowing the risks she was taking, Buckmaster [chief of French Section] ordered her to return to London but she refused to leave until she was satisfied that he'd found a replacement for her. He assured her that he had, and she finally agreed to be picked up by Lysander in mid-October. She then went off the air for ten days, and missed the [full] moon period [for scheduled pickups]. She surfaced again on 18 October with a new batch of messages, and although their security checks were correct the first message had a transposition-key eighteen letters long.

In fact, having been betrayed by someone in her network, she'd already been in Gestapo custody since sometime around 13 October.

Marks understood the dire import of this message — one perhaps sent by an imposter telegrapher. But he did so only by going outside SOP to double-check:[114]

> I hadn't realized the extent of the carelessness [among the decoding staff] ... until Noor's message to London with eighteen letters in her transposition-key. I'd left strict

---
112 This rule was inspired by a similar line by Laura Lippman, *What the Dead Know*. (2007), 42.
113 Marks (1998), 399 and 511 for a potential backfire when using this kind of "special check".
114 Marks (1998), 419.

instructions with the station that, in the event of this happening, the supervisor was to notify me at once and teleprint the code-groups to [me]. But the supervisor had taken no such action, and if it hadn't been for my unease about Noor I wouldn't have asked to examine the code-groups. I then discovered that she'd inserted her distress signal.

Marks says that when reporting this to Colonel Buckmaster that: "This confirmed his [B's] suspicions that she was in enemy hands as the style of her new messages had changed."[115] Although SOE's French Section had made and would continue to make many blunders,[116] they never achieved the stratospheric level of incompetence that marked all but the final change of command at Dutch Section (see CASE 4.8.2). All of Dutch Section's agents & cells ("circuits") had come under German control, while only a fraction of French Section's had.

Incidentally, I read this large discrepancy in incompetencies between the Dutch & French sections as nearly conclusive evidence against the still rampant conspiracy theory that SOE had been deliberately sacrificing its agents (particularly the women) for some devious reason — such as to protect such major secrets as, variously suggested, the Allied invasion of Sicily (Jul 1943) or of France (Jun 1944).[117] Had that been the aim of the "perfidious" English, I'd expect to find the reverse or at least equal proportion of slip-ups because France was far better placed on Hitler's strategic map for playing such tricks.

With Allied armies closing upon the German frontiers, it became Nazi policy to destroy evidence of such secret matters. Noor was taken to one of the Nazi extermination camps, almost certainly Dachau where sometime in September she was executed. But the Nazis weren't through with her. The final indignity was in the post-war testimony of her captor in Paris, SS Major Hans Josef Kieffer, that during her lengthy interrogation, while Noor had given up no secrets — not even her real name — she'd repeatedly *lied*. The mute testimony

---

115 Marks (1998), 399. Conversely, Foot (2002), 301, writes "Marks went to Buckmaster with the text, as proof; Buckmaster did not believe him." Foot cites both Marks (1998), but only pp.200-201, which are not relevant to B's beliefs; plus (p.497n189) a conversation with Marks in 1998, which I assume is his basis of this quote. As Marks had contempt for B and as B had been dead 6 years when M published, it would seem unlikely that he would have protected B in this matter. However, we now know that Her Majesty's Government had held up publication of M's manuscript, which he'd finished by 1988, and B didn't die until 1992. So, while I'm quite prepared to believe the worst of Colonel B, I've been unable to resolve these versions.

116 A detailed catalog & analysis of the blunders by SOE, its agents, and the French Resistance is Foot (2002), Chapter X ("A Run of Errors: 1943-1944"), 258-308.

117 This unlikely but persistent conspiracy theory has been best refuted by Foot (2004), 273-276, where he explains (p.275): "The truth is that PROSPER's downfall, tragic as its consequences were, was brought on in spite of their bravery by the agent's own incompetence and insecurity." See also M. R. D. Foot, "Research Note: The Dutch Affair," *Intelligence and National Security*, Vol.20, No. 2 (Jun 2005), 341-343.

of her "distress check" suggests that, true to her Sufi belief, she'd spoken only truth. Just not the whole truth.

Here was a rare case of a compulsive truthteller letting those deceive themselves who would have used her in their flim-flam game of double-agentry. By withholding the crucial secret of her distress code that Leo Marks had given her, Noor had used this weapon to cause the enemy to unwittingly break their own game.

5. The Security of Options

> A further touch of artistry in deception is to provide an alternative to your true intentions so valid that if your adversary detects it as a hoax, you can then switch to it as your major plan and exploit the fact that he has discounted it as a serious operation
>
> — R. V. Jones, "Intelligence and Deception" (1981), 19

Stratagemic security is absolute — but only when the deception operation succeeds in anticipating the preconceptions of the victim and playing upon them. In such cases the victims becomes the unwitting agents of their own surprise, and no amount of warning (i.e., security leaks) will suffice to reverse their fatally false expectations.[118]

Deception, by its very nature, provides its own best security. Although the term "security of options" was my original coinage,[119] the concept was only my generalized extension of Liddell Hart's principle of "alternative objectives", first formulated by him in 1929.[120] Alternative objectives means simply that every football play, most military offensives, and many other operations, can have two (or more) goals. Wherever possible, we should be prepared to switch to a second-best goal rather than continuing to batter away in mounting frustration at our primary target. Or even adopt that second-best option at the outset in those all too common situations where our opponent clearly expects and has prepared to contest the obvious first choice.

Whenever a deception operation succeeds in anticipating the preconceptions of the victim and playing upon them, deception security is absolute. Then the victim becomes the unwitting agent of its own surprise, and no amount of warning (such as security leaks) will suffice to reverse it fatally false expectations.

The worst possible case would occur if the deception plan itself were prematurely disclosed to the victim (without the deceiver knowing). While

---

118 For example Whaley, *Codeword BARBAROSSA* (1973).
119 First stated in Whaley, *Stratagem* (1969), Chapter 6.
120 The concept of "alternative goals" or "alternative objectives" was first stated by B. H. Liddell Hart in his *The Decisive Wars of History: A Study in Strategy* (London: Bell, 1929), 141-158. See also his *The British Way in Warfare* (London: Faber & Faber, 1932), 302, etc; *Strategy: The Indirect Approach* (New York; Praeger, 1954), 161-164, 333-372; and *Memoirs*, Vol.I (London: Cassell, 1965), 166-168. See discussion of the evolution of this concept in Whaley, *Stratagem* (1969), 128-139.

such potentially disastrous disclosure is rare in general and so far unknown in war.[121] The closest to this was during the Battle of Midway in 1942 (Case 4.6.4, above) when faulty Japanese security permitted the U.S. Navy to see *around* the primitive Japanese deception operation and set an ambush.

Even if the deception plan was compromised, all wouldn't necessarily be lost. First, the disclosure itself would have to be believed. Second, if the deceiver knows or even suspects disclosure, she can actually capitalize on this by switching to one of the alternative courses of action or simply adopting a new deception plan to reverse appearances, as British General Alexander did in breaching the German Gothic Line in Italy in 1944 (Case 4.7.5, above) by deliberately disclosing an already successful deception operation and substituting a different one. Even if the direction or objective of the attack has been compromised, the planner can still manipulate the victim's perception of the timing or strength or even the intent or style of the attack.

Even if the deception plan runs counter to or fails to play upon the victim's preconceptions, the very fact that it threatens alternative objectives will usually assure enough uncertainty to delay or diffuse or otherwise blunt the victim's response.[122] The worst possible case would occur if the deception plan itself were prematurely disclosed to the victim. While I'm unaware of a military example,[123] even if it did, all would not necessarily be lost. First, the disclosure itself would have to be believed. Second, if the deceiver knows or even suspects disclosure, he can actually capitalize on this by switching to one of the alternative courses of action or simply adopting a new deception plan to reverse appearances.[124] Finally, even if the specific *direction or objective* of the attack has been compromised, the planner can still manipulate the victim's perception of the attack's timing or strength or even its intent, payoff, or style — indeed of any of the Nine Categories of things about which one can be deceived or surprised.[125]

---

121 A recent example from strategic international politics was the disclosure in 1986 of the CIA's disinformation campaign against Libyan leader Gadhafi. At less exalted levels we see this in "tip-offs" of con games and other scams and commonly in "poison-pen" letters that reveal marital or other infidelities.

122 Examples include Whaley, *Stratagem* (1969/2007), Cases A49 (Leyte) and A65 (Bay of Pigs).

123 The closest example I've identified is in Whaley, *Stratagem* (1969/2007), Case A33 (Midway), when faulty Japanese security let American naval intelligencers see around the primitive Japanese deception operation and set an ambush.

124 An instructive analogy is provided by Whaley, *Stratagem* (1973/2007), Example B38 (Gothic Line) where British Field-Marshal Alexander successfully reversed an already successful deception operation by deliberately disclosing the first one.

125 The 9 categories are: CHANNEL, PATTERN, PLAYERS, INTENTION, PAYOFF, PLACE, TIME, STRENGTH, and STYLE. See Whaley, Deception Verification in Depth & Across Topics (FDDC, Nov 2009), 30.

CASE 5.1:
The No M.O.

> "You know my method. It is founded on the observation of trifles."
>
> — Sherlock Holmes to Dr. Watson in "The Boscombe Valley Mystery," *The Strand Magazine* (Oct 1891)

Whenever we learn an opponent's *modus operandi* or mode of operating (M.O.) we become able to efficiently prioritize our analysis of what that person or organization may be thinking, plotting, or doing.

The earliest explicit fictitious version of this notion was the basis of a short story published in 1908 by a brilliant young British military innovator (co-inventor of the tank, among other things). Ernest Swinton's plot had his hero, an army chief of staff, stalled on the proverbial horns of a dilemma over which course of action his unknown opposing general would chose in his imminent attack. Only when the hero learns that officer's identity — a man who'd tricked him before — does he know his enemy's M.O. and hence which winning strategy to recommend to his own commander

A real-world example involved Dr. R. V. Jones, the WW2 chief of British Air Ministry scientific intelligence. He had an old friend in Germany who'd been his near equal as both a physicist and as a practical joker. This deceptive scientist, Carl Bosch Jr, was worthy of concern. As Jones wrote:[126]

> Sometimes during the war I had wondered what he was doing: was he my "opposite number"? If so, he would know all my weak points and he was such an excellent hoaxer that he might easily have misled us.

As Jones would only learn 40 years later (when he met Bosch in Miami where the German was working for NASA), Bosch had in fact worked on two of the same projects (centimetric radar and radio beams for rockets). Fortunately, because the Germans underplayed scientific intelligence, Bosch had been only intermittently and part-time assigned to such vital countermeasures.

Incidentally, this potential for a Jones-vs-Bosch type of battle-of-wits suggests a need for the systematic collection & analysis of biographical intelligence on personnel specifically engaged in political-military deception only of somewhat lower priority than the usual biographical profiling & gaming of opposing national leaders and military commanders.

---

126 Jones (1978), 502.

Jones needn't have been overly worried that Bosch or any other opponent would know enough about him that they could outplay him by being that proverbial one step ahead in a game of wits. Although Bosch would have known some of Jones's "weak points", he wouldn't know all of Jones's M.O. That was because Jones had no *fixed* M.O., no fixed style, that could betray him. Of course, just knowing that any person like Jones was a world-class practical joker greatly helps to anticipate some of the general types of tricks he might be planning. But, unless we read his mind, we can't predict what other tricks he might try.

CASE 5.2:
The Secret M.O.

Similarly, most professional stage magicians have no fixed M.O. They are known to be more-or-less skilled in the three main branches of conjuring: sleight-of-hand, apparatus magic, and mental magic or mentalism. By switching between or combining these styles they make detection of their method for certain tricks more difficult.

Conversely, one master magician managed to thoroughly deceive other master magicians for decades by having a secret method. This was the great British conjuror, "Cardini" (Richard Valentine Pitchford), who performed professionally from 1918 to 1966. World renowned for his sleight-of-hand with card, cigarettes, and other small objects, that was his signature style, his public M.O. Known only to a few trusted colleagues, he also had a secret M.O. For a few special tricks he would use a concealed mechanical reel of "invisible" thread.[127]

Because his very name and billing as "Cardini the Suave Deceiver" had come to be perceived as a pure sleight-of-hand artist, his occasional use of apparatus went undetected by even top experts. They'd been conditioned to exclude apparatus from the range of possile hypotheses about how any given trick of his might have been done.

CASE 5.3:
Sherman's March to Atlanta, 1864

In 1864 Yankee General William Tecumseh Sherman made his decisive 180-mile drive through Georgia to Atlanta (which with his 300-mile follow-through to the sea sliced the Confederacy in two).

Throughout his approach to Atlanta, Sherman's logistic tail was tied to a single railway line. He had to advance and attack along that line, a fact that the Confederates knew and that he knew they knew. Yet in every engagement

---
127 Conversation with magic historian Jeff Busby, 1992.

but one — a costly frontal attack at Kennesaw Mountain — he surprised the awaiting defenders as to the *place* of his attack, defeating them each time. How was this possible?[128]

Only one option remained but Sherman exploited it to the full — the old left/right option. Moreover, he was forced to rely on this two-pronged option repeatedly throughout this campaign, which thereby became that quite rare process, what I've call a campaign of "serial deceptions". Although his line of advance was narrowly constrained, he retained at the spearhead the alternative of attacking either to the right on the railway line or to the left. He literally, as he wrote to General Grant at the time, placed his enemy on the "horns of a dilemma".[129] Right flank or left, he always succeeded in using tactical deception to conceal which it would be.[130] And having done so he then, to use Liddell Hart's imagery, impaled Johnny Reb on the chosen horn. Interestingly, while Liddell Hart modestly credits Sherman with this consequent forced-choice,[131] the explicit statement of the concept is original with Liddell Hart. Sherman only very ambiguously implies it. He'd either not fully understood his own great insight or failed to state it.

This is a spectacular example of *serial* induction of misperception of place.[132] Serial deception involves what Liddell Hart, borrowing a phrase from Rugby football, called "selling the 'dummy' first one way and then the other".[133]

CASE 5.4:
The Two Patton's Ruse, 1944

Urgent improvisation of a deception plan in its extreme form was thrust upon Ralph Ingersoll during the critical Battle of the Bulge. It was December 1944 and the German army had just broken through the Allies' weakest point in a surprise attack. One measure of this surprise was that Col. William "Billy" Harris was away on other matters. Harris headed Gen. Omar Bradley's 12th U.S. Army Group's "Special [i.e., deception] Plans Section". Thus, at this crucial moment, command devolved upon Major Ingersoll.[134]

On December 19th General Eisenhower went to Bradley's rear HQ at Verdun to meet Bradley and Patton and decide on a strategy to stop the German

---

128 Whaley, *Practice to Deceive: Case Studies of Deception Planners* (draft, 1987).
129 Sherman to Grant, letter dated 20 September 1864, in William T. Sherman, *Memoirs* (Bloomington: Indiana University Press, 1957), 115.
130 Liddell Hart (1954), 149-153; and Whaley (1969), 129, 136-137.
131 B. H. Liddell Hart, *Sherman* (New York: Dodd, Mead, 1929), 315-316.
132 General Alexander's Italian Campaign of 1943-45 is another good example of serial deception.
133 A general discussion with other examples and references is in Whaley (1969), 136-139.
134 Whaley, *Practice to Deceive: Case Studies of Deception Planners* (draft, 1987); BW's interview with Ralph Ingersoll, 12 May 1973; and also Roy Hoopes, *Ralph Ingersoll: A Biography* (New York: Atheneum, 1985), 304.

offensive. Ingersoll was present at the momentous 11 A.M. meeting. The battle-eager Lt. Gen. Patton assured Eisenhower and Bradley that he could rush his powerful Third Army north and stop the enemy advance. These two senior commanders were doubtful because of the immense logistical problem, but desperate measures were needed. So they unleashed Patton.

Everyone involved knew it was too late to mount any conventional deception operation to cover this counterattack. Nevertheless, General Bradley turned to Major Ingersoll and asked "Is there anything you and your people can think of to throw the Germans off – about where they [Patton's 14 divisions] are going to strike?" Ingersoll replied "Yes, sir. We can do" – he hesitated – "something." Bradley was momentarily silent before simply ordering "Then do it."[135]

Ingersoll later told me and, still later, his biographer that the idea had come to him in a flash. To me he explained what had inspired him. Realizing during the meeting that time would be too short for any conventional deception, he had been trying to think of some alternative. He was particularly mindful of one overriding constraint — urgency demanded that Patton's movement orders to his corps and division commanders would have to go out over his radio nets "in clear" (uncoded), despite the certainty that the German signals intelligence teams would overhear.

Ingersoll's mind lit on the recent example of the vast confusion to German military intelligence engendered on D-Day by the unintentional scattering of the two American and British airborne divisions. With this thought in mind, Ingersoll suddenly conceived — indeed invented — the unprecedented ruse of The Two Pattons.[136] He personally codenamed the operation KODAK because he intended that it confuse the enemy by giving them a "double-exposure".[137]

The only resource available to Ingersoll was Special Plans' operational deception unit, the 23rd Headquarters Special Troops. It was too late to deploy their usual panoply of visual and aural spoofs. Radio deception working alone would have to do the job. Accordingly 19 officers and 20 enlisted men were detached from the 23rd's Signal Company, rushed forward, and dispersed among advance units of Patton's Third Army. An additional 11 officers and 205 EM provided support further back. It was, as the official history called it, "a big (29 sets), loose radio show".[138]

These radio operators normally spent their time simulating fictitious U.S. Army units in a largely successful effort to dupe the enemy to waste precious

---

135 Hoopes (1985), 304.
136 Elsewhere I had dubbed this the "Five Patton's Ruse", an error corrected by reading the recently declassified official history of the 23rd Hq Special Troops. In my interview with Ingersoll he had, as Went Eldridge warned me, exaggerated, thereby inventing an even more dazzling illusion. See Whaley & Bell (1982), 379-381.
137 *Official History of the 23rd Headquarters Special Troops* (1945), 23.
138 *Official History 23rd* (1945), 23, and also Enclosures I and V.

resources by either advancing or retreating from these phantoms. But now, for two days, December 22nd and 23rd, Ingersoll had them imitate two real units – the 80th Infantry Division and the 4th Armored Division that were spearheading Patton's advance. The 23rd bogus radio show was timed to start when these divisions' radio nets began broadcasting "in clear".

The real 80th was getting ready to jump off due north from Luxembourg, and the real 4th was preparing to roll up from Arlon to effect its historic relief of the 101st Airborne Division trapped in Bastogne. "The 23rd mission was to show ... the presence of the 80th Infantry and 4th Armored Divisions slightly northeast of Luxembourg and in position to forestall any German plans of extending their counterattack southwest through Echternach."[139]

In effect they presented German intelligence with two General Pattons, each approaching the Bulge from slightly different directions. Of course, the Germans quickly understood the radio game being played against them, but they were still dazzled by the two fake divisions advancing eastward of the two real ones. In the event, the Germans were able to keep track of only one of Patton's 14 advancing divisions. All others were either lost entirely on the German battle maps or, worse, mislocated. Consequently Patton was able to gain the substantial tactical surprise that helped him break the back of the Wehrmacht's last offensive.

Two decades after the war, I interviewed Ingersoll about his role in military deception. The Two Pattons story — unpublished at that time — emerged only as an afterthought when Ingersoll explained apologetically that, while proud of his improvisation, he didn't class it as "deception", which he defined as making the victim certain but wrong. Instead he chose to make the enemy merely uncertain, confused, and hope for the best. He was too modest. In fact, Ingersoll's unique ruse does indeed fit one category of deception, the one called "Dazzle" in the Bell-Whaley typology.[140] True, it was a desperate measure provoked by a desperate situation, precisely the type of rare situation where magician Rick Johnsson's Too Perfect Corollary applies.[141]

---

139 *Official History 23rd* (1945), 23.
140 Whaley (1982), 184.
141 Rick Johnsson, "The 'Too Perfect' Theory," *Hierophant*, No.5/6 (Fall/Spring 1970/71), 247-250.

6. Double or Not?: Two Parallel Cases

Both following cases are presented as parallels. The first is a thoroughly documented case of double agentry. This second is a persistent unresolved mystery — was he or wasn't he a double agent. The first is a British case, the second an Israeli one — one that echoes the infamous American cases of Nosenko and Loginov.[142] To help the reader evaluate each case while partly nullifying the 20/20 hindsight effect, the evidence is presented in timelines.

These two cases could have been included with the other cases of double-agentry in Section 4.8. But I've chosen to separate them from the others and place them together here to better demonstrate one simple but effective way to notice and focus on the incongruities that are inherent in any deception. This method is what I've called Timeline Deception Analysis (TDA). It has three distinctive components.

- It displays the data chronologically. This, of course, is simply the first step in the so-called Historical Method. But it's particularly useful for analyzing past cases where deception is suspected because it's the single most effective way to break the dead grip of hindsight knowledge, enabling the analyst to revisit old evidence and perceive new hypotheses.

- In the short term, it keeps us alert for Red Flags, those often obvious, even flaunted, signals that both the trickster and the agent must use to obtain the cooperation of their target victims in supplying either material goods (the con artist) or information (the double agent). Red Flagging is, of course, essentially only a form of "profiling." As such it has profiling's value of giving a "head's up" to put us on higher alert. But it also has profiling's enormous weakness of *always* being inconclusive.

- For verification and for the longer term, it will also use systematic Incongruity Analysis, which is the only sure way of conclusively proving deception.

These three components work together synergistically as an efficient and often rapid method for detecting deception — both in general and particularly when the deceiver is being double-crossed.

---

142 See Whaley, *Deception Verification in Depth & across Topics* (FDDC, Nov 2009), Chapter 3.3; and Heuer (1987).

CASE 6.1:
GARBO, 1944

> This was an old trick ... in which you give your victim something genuine that you know he already knows, in the hope of convincing him that the rest of the report, which contains the hoax, is genuine.
>
> — R. V. Jones, *The Wizard War* (1978), 70

Juan Pujol was an anti-Nazi Spaniard. But he'd acquired plausibly right-wing credentials in the late 1930s when he'd deserted from the leftist government's army to Franco's rebel fascist army during the Spanish Civil War. Capitalizing on that credibility and hoping to subvert from within he obtained employment in 1941 with German Military Intelligence (the Abwehr) as one of their agents (codenamed ARABEL). Then in 1942 he convinced the British Double-Cross Committee in London to take him on as a double agent (codenamed GARBO). Over time he and his British case officer, Tomás Harris, got the Abwehr to believe he'd built up his own GARBO NETWORK through Britain of 27 sub-agents, all of whom were, in fact, fictitious. As that infamous agent-in-place, Kim Philby, remarked in collegial admiration, his friend Harris was about to "conceive and guide one of the most creative intelligence operations of all time."[143]

Now, let's focus on two timelines. The first GARBO timeline is set in 1942 and concerns the Allied amphibious landings in North Africa:

TORCH D-minus-10 (29 Oct 1942)

> The Allied landings in North Africa were codenamed TORCH. D-Day was scheduled for 8 Nov 1942. The Double-Cross Committee decides that GARBO will send a slow letter to his Abwehr controller giving early warning of the landings — early enough to seemingly confirm his genuineness (*bona fides*) but too late for the German military to respond effectively. REFERENCES: Harris (2000), 103; Crowdy (2008), 163.

TORCH D-7   GARBO sends that warning. REF: Harris (2000), 103.

TORCH D-1   GARBO's follow-up and more urgent & detailed warning released. Sent air mail to Lisbon for circuitous forwarding through Madrid where it would be radioed to Berlin. SOURCE: Harris (2000), 103-104 and 105 for details of the

---
143 Kim Philby, *MySilent War* (New York: Grove Press, 1968), 19.

communication system involved. Incidentally Pugol (1985) himself later entirely overlooked his time-delay trick in TORCH, further proof that he wasn't its inventor.

NOTE: The Abwehr and the German High Command now judged GARBO and his network to be one of their best and most trustworthy sources of high quality and timely intelligence. This was still the situation in May 1944 (Normandy D-Day minus 23) when his British controllers began to intensify their use of him as their main channel for feeding the Germans on FORTITUDE-South, the chief Channel D-Day deception that falsely portrayed the Pas de Calais as the target of the main Allied landings instead of the "diversionary" landings 180 miles westward in Normandy.[144] Since the beginning of the year GARBO has sent more than 500 wireless message (averaging 4 per day) to his receiver in Madrid. REF: Cowdry (2008), 261. The timeline resumes:

FORTITUDE D-minus-31 (1944, 4 May)

    Harris concocts the it's-later-than-you-think D-Day ploy for GARBO to use immediately before the invasion. REFS: Delmer (1971), Chapter 10 ("Madrid's D-day Bungle"), 174-191; Pujol (1985), 130, including explicitly crediting Harris with this timing ruse; Holt (2004), 146, 577. Delmer's is the most detailed published account of this very special stratagem.

FORTITUDE D-23

    GARBO starts a series of reports indicating that the main Allied cross-Channel attack would be preceded by several diversionary landings. It was gradually suggested that the main *diversion* would be a landing on the Normandy beaches (which, of course, was all along the real Allied target). Then, days or even weeks later, the "real" invasion would hit the beaches at the Pas de Calais. The size of Allied forces needed for two such major landings was only plausible to the Abwehr because the double-cross agents had already sold the idea that the Allies had twice as many troops available in or on their way to Britain than was the case. REF: Harris (2000), 194.

D-7 to D-4   GARBO sends series of messages to enhance the Germans' perception of a threat to the Pas de Palais (from Patton's notional FUSAG). REF: Harris (2000), 197.

    This ruse is plausible because the Germans had already (by 15 May) accepted the British disinformation (including from GARBO) that the Allies have *twice* the number of troops and

---

144 Tomás Harris, *GARBO: The Spy Who Saved D-Day* (Richmond: Public Record Office, 2000), 192-241.

	combat units already in Britain or due to arrive before D-Day. This false order-of-battle encourages the belief that the Allies have the strength to hit more then one place at the same time. REFS: Harris (2000), 192-194, 201-202; Masterman (1972), 155-157.
D-3	Madrid Abwehr tells GARBO that his coded wireless messages from London will take about 48 hours to be read in Berlin HQ. SOURCE: REFS: Harris (2000), 197; Pujol (1985), 130.
	Now, knowing the exact degree of time delay for warnings, Harris can carry out the time trick he'd conceived on D-minus-31 by giving the Abwehr 3½ hours warning, thereby gaining further credibility for GARBO with the Abwehr.
D-2	GARBO's (notional) agent in Scotland gets a codeword that he's to immediately *telephone* to London on first indications that the Scotland-based portion of the Allied invasion force was embarking from Glasgow. "This," as Harris wrote, "was to ... keep the Germans in Madrid on the alert on the eve of D Day so that we should be enabled to pass them a last moment [H-3½ hour] pre-release that the invasion was about to commence." REFS: Harris (2000), 197-198; Pujol (1985), 133; Holt (2004), 557; Crowdy (2008), 265-266.
D-1 (4 Jun)	GARBO ready early next morning (the 5th) to radio his Madrid controller warning of the imminent landings in northen France. REFS: Pujol (1985), 134; Harris (2000), 26, 197-198; Delmer (1971), 179-180.
	However, bad weather in the English Channel on the 4th causes SHAEF (Eisenhower) to order the D-Day clock reset 24 hours ahead to the 6th. Consequently, GARBO's radio warning message is also advanced 24 hours.
REVISED D-1 (5 Jun)	
	GARBO, forced into a 24-hour delaying mode, radios only redundant information to Abwehr- Madrid which, per GARBO's request, had stayed open during the night. REFS: Pujol (1985), 133-134; Harris (2000), 198.
D-Day, H-minus 3½ hours	
	Abwehr-Madrid, to GARBO's disappointed surprise, fails to stay on the air and thereby misses his 3½ hour advance warning, which he transmits repeated from around 0300 hours

on. REFS: Pujol (1985), 130, 135; Delmer (1971), 180-181; Harris (2000), 198; Crowdy (2008), 265-266.

D-Day, H-Hour

Before dawn on 6 Jun at 0630 hours the Allied invasion armada is visually verified by the surprised enemy troops on four Normandy beaches to be approaching and beginning to off-load. Paratroops had already been dropped inland.

D+1 GARBO informed (as he'd already known was the case) by Abwehr-Madrid that his crucial invasion warning had not been received much less forwarded. He rages (at 2155 hours and again on D+6) at Abwehr-Madrid having gone off the air contrary to his advice. Madrid apologizes. REFS: Pujol (1985), 140-141; Harris (2000), 204; Crowdy (2008), 268.

D+2 GARBO claims that day-before-yesterday's Normandy landing was only a "diversionary manœuvre designed to draw off the enemy [German] reserves in order then to make a decisive attack at another place." The next day the German High Command (OKW) underlines that passage and comments that it "confirms the view already held that a further attack is to be expected in another place (Belgium?)." REFS: Pujol (1985), 142-146; Hesketh (1999), xxiii-xxv, 208-210; Harris (2000), 205; Rankin (2008), 402, 407; Crowdy (2008), 270-272.

In sum, even long after the event, German intelligence and the high command remained convinced that GARBO had been their controlled agent and not a British double agent. I'm surprised they'd failed to question the coincidence that both of GARBO's most potentially consequential warnings (on TORCH & FORTITUDE) had while (notionally) *sent* in timely fashion, did not *arrive* at German GHQ until too late for action. The Germans blamed only their own slow communications and failed to even consider that GARBO had been manipulating that weakness.

CASE 6.2:
Marwan, 1973-2007

> I admire the ingenuity of your hypothesis, but ....
>
> — R. V. Jones, letter to David Irving, 27 Mar 1984, in response to yet another of Irving's dotty notions.

The 1973 Yom Kippur surprise attack by Egypt (and Syria) was the most devastating event in the history of modern Israel. A key figure had been Dr. Ashraf Marwan, then a senior Egyptian official who was either an Israeli espionage agent working against Egypt or an Egyptian double agent working against Israel. His case is a challenging parallel to the Nosenko & Loginov cases — one deserving close analysis along the lines of Heuer (1987) — a much closer analysis than I've been able to apply here.[145] But it also bears a suspicious resemblance to the GARBO time distortion game just described above.

Marwan's timeline begins:

1944, 2 Feb	Mohamed Ashraf Marwan born in Cairo. His father was a military officer who later served in President Nasser's Presidential Guard.
1965	BSc in Chemistry from Cairo University. At this time he has little money and only modest prospects. Joins the Egyptian Army.
1966, Jul	Becomes son-in-law of Egyptian President Gamal Abdel Nasser upon marrying his younger daughter, Mona (Muna), whom he'd met & pursued at Cairo University. Begins work in a senior position with Nasser's Presidential Information Bureau (essentially a central intelligence service) under its pro-Moscow head, Sami Sharaf.
1968	Comes to London to study for his MSc in Chemistry. Although it is known that Marwan has a taste for high living, it is rumored than he is already growing wealthy from his official positions — rumored as 5% of all funds transferred from the Arabian peninsula.
1969	Walks into a Mossad front in London, a physician's office, where he drops off his X-rays, a large file of top secret Egyptian documents, & offers his services. Three days later, after checking as best they can, Mossad recruits him.

Marwan periodically meets his Mossad handlers in a London safe house where he demands & is paid a lavish fee (variously reported as US$100,000, £50,000, & £100,000) per meet. These meetings, which often included Mossad Director Zvi Zamir, were taped and transcribed for the Israeli Prime Minister. The Israelis begin to call Marwan "The Source". Over the next 4 years Mossad reportedly paid him US$3 million (another report gives the unlikely sum of US$20 million). |

---

145 Bar-Joseph (2008); *Wikipedia*, "Ashraf Marwan" (accessed 1 Jan 2010).

1970, Jan	In Moscow President Nasser repeats his request for long-range fighter-bombers that he believes Egypt needs to knock out the Israeli air force as a prerequisite for an Egyptian invasion. Marwan gives Mossad a transcript of this conversation. This intelligence is a key part of the thence Israeli SOP theory ("The Concept") that Egypt will never attack unless these two preconditions (fighter-bombers & scud missile) come about.
1970, Sep	Nasser dies and is succeeded as president by Anwar Sadat. Marwan serves Sadat initially as an Adviser for Arab Affairs (a kind of roving ambassador) while continuing under Sami Sharaf at the Presidential Information Bureau.
1971, Feb	Sadat indicates (in letter to UN mediator Jarring) that Israel can have peace if it will return the territories occupied after the 1967 Six-Day War. Israel rejects this initiative (which was not entirely sincere, if at all).
1971	M allegedly volunteers to work for Italian intelligence, a relationship that continues at least until the Yom Kippur War.
1971	Marwan approaches the CIA COS-Egypt, reporting on conspiracies against President Sadat. Media later allege he was a CIA recruit; but it sounds more like he recruited them.
	At some point M is spotted by Mossad officer Peter Malkin driving to one of his Mossad meets in London in a car bearing Egyptian licence plates — a potentially fatal breach of tradecraft for an Egyptian spy for Israel but not for a double agent.
1971, May	M, having helped foil an anti-Sadat political conspiracy (whose leaders included Sami Sharaf), is promoted by Sadat to succeed Sharaf as head of the Presidential Information Bureau. Incidentally, this brings a shift away from the pro-Moscow leaning of the failed coup leaders to Sadat's more neutral & nationalistic stance.
1971	M begins writing his memoirs — at least that's what Israeli-British historian Dr. Ahron "Ronnie" Bregman would tell *Haaretz* in 2007 as M had told him.
1972, Feb	Masterman's *The Double-Cross System* published. Includes a vague reference to the D-Day time delay ruse (see CASE 6.1, above). A best-seller, I assume Zeira would have read it and, as a professional deceptionist & amateur magician, have had the "prepared mind" to see its significance.

1972, early	Sadat publicly announces Egypt's commitment to wage a war of extermination on Israel — including, if necessary, the commitment (death?) of a million Egyptian soldiers, an Armageddon-scale threat not lost on the Israeli leadership. Egypt begins a major drive for armaments.
1972, Aug	President Sadat messages Soviet President Brezhnev that Egypt needs Scud retaliatory missiles as an additional prerequisite of any Egyptian attack on Israel. Marwan also passes this document to the Mossad, thereby solidifying the official Israeli theory of "The Concept".
1972, Oct	Sadat secretly drops his former war policy of awaiting special weapons (fighter planes & missiles) to one of attacking without them. Moreover, his new war goal is much more modest — not total but limited war. Marwan knows about this crucial policy change but may not have reported it to Mossad (although Bar-Joseph claims he did).
1972, Dec	Marwan, using the codeword RADISH meaning "imminent war," brings Zamir by plane to London where Marwan elaborates but, by not specifying a day or hour, Israel goes on alert but doesn't call up the reserves. It proves a false alarm that starts to give Marwan a reputation for "crying wolf".
1973	M receives PhD in Organic Chemistry from Cairo University. [BW: Not "a PhD in Economics in Britain" as misstated by *Wikipedia* [accessed 13 Jan 2010].
1973, Apr	M again warns Mossad that Egypt & Syria will launch a surprise attack on Israel on 15 [sic] May. Israel accordingly goes on high alert — an expensive false alarm — in spite of Gen. Zeira's advice that Egypt didn't intended to attack. This consolidates Marwan's "cry wolf" reputation — and Zeira's reputation for caution. Bar-Joseph claims this was a real warning, the date having been put off only because Syria wasn't ready.
1973, c23 Aug	A meeting of the joint Syrian-Egyptian Armed Forces Supreme Council agrees on the possible dates of their combined attack on Israel.
1973, 28 Aug	Sadat flies to Saudi Arabia where he informs Saudi King Faisal of war on Israeli in the very near future, according to British author Jeffrey Robinson's *Yamani* (1988). Because R mentioned that his Marwan was the only other person present at that meeting, Zeira (writing in 1993 in his *Myth versus*

Reality) will conclude that, because Marwan did not pass this warning to Mossad, he must have been a double agent. Indeed, although Marwan reports this meeting to Mossad, he lies that Sadat had later decided to postpone the war.

1973, 5 Sep By arresting Palestinian terrorists and seizing their SAM-7 missile, Italian police thwart a Palestinian operation to shoot down an El Al passenger plane at Rome Airport.

In late 2002 Israeli-British historian Dr. Ahron "Ronnie" Bregman reveals in his new book that this had been a joint Libyan-Egyptian plot in reprisal for Israeli IAF fighters having shot down a Libyan passenger plane that had accidentally strayed over Israeli-occupied Sinai in Feb 1973. Bregman further discloses that this plot had been handled from the Egyptian side by Marwan (who personally supplied the missile) and that M was also the informant who'd tipped off the Italian police. This secret betrayal of Libyan President Qaddafi enabled Sadat to avoid inflaming Israel while still seeming to be Qaddafi's friend as well as restore Israeli confidence in Marwan after his twice crying wolf. REF: Ahron Bregman, *A History of Israel* (UK: Palgrave, 2002), 145-150.

1973, 25 Sep Jordan's King Hussain personally warns Prime Minister Golda Meir that the Syrian army was already in its "pre-jump positions" for a joint Syrian-Egyptian surprise attack on Israel. The Israeli leadership, disbelieving, takes no action.

D-2, 4 Oct (Thursday)

Marwan flies from Egypt to London.

That day, during a meeting in Damascus, Egyptian War Minister Ahmed Ismail & Syrian President Asad agree to advance H-Hour four hours to 1400 hours. According to Bar-Joseph, Marwan could not know of this change because he'd already left Egypt to go to London.

D-1, 5 Oct 1973 (Friday morning)

Some 40 hours before Yom Kippur, Marwan's Israeli case officer ("Dubi") in London telephones Mossad chief Zamir in Israel. He reports that Marwan requests a meeting in London that day. Zamir assumes the matter is serious because Marwan had again used the codeword RADISH meaning "imminent war".

	Zamir, moving slowly, takes the first El Al commercial morning flight to London. Late that night (near midnight) at a safe house in London Marwan (who arrived late) tells Zamir that Egypt & Syria will attack at next day's sunset, that is, around 1800 hours. Because the Israeli Embassy communication center is closed [!!!] for the Yom Kippur holy days Zamir transmits the warning by public telephone using improvised codewords. [BW: Can this communication bumbling be even half true?]
	Israelis learn that Egyptian soldiers have been ordered to break their Ramadan fast — an unprecedented order indicating something serious was afoot.
D-Day, Saturday	
	Delayed by the slow holiday answering of Israeli telephone operators, Zamir's call doesn't get through to Mossad HQ until 0340 and thence to Israeli decision makers until shortly before dawn. The warning isn't credited by Israeli Military Intelligence (Zeira). IDF Chief of Staff Elazar is notified at 0430.
	At 0805 Prime Minister Meir, Defence Minister Dayan, and Chief of Staff Elazar meet and decide to do nothing other than move some armored units forward. Consequently, mobilization of the reserve, which is two-thirds of the Israeli Army, was not ordered. (The Egyptians assumed it would take the Israelis 5 days to fully mobilize.)
D-Day, H-Hour	
	On Saturday at 1400 hours, 6 Oct 1973, Egypt launches its attack 4 full hours before Marwan's promised H-Hour of 1800 hours.
1974, 1 Apr	The Israeli Agranat Commission's Interim Report on the Yom Kippur War released. Without naming Marwan it refers to him as "The Source". Naming them as responsible for the surprise, the report calls for and receives the resignation of Chief of Staff Elazar, Chief of Military Intelligence Zeira, and Chief of the Southern Command Gonen. Public outrage forces the resignations of Prime Minister Meir that month and Defense Minister Dayan the next.
1975, May	M becomes founding Chairman of the Board of the Arab Organization for Industrialization (AMIO, later AOI), an Egyptian government-sponsored consortium of Arab states to promote high-tech weapons production in Egypt with

	sales abroad, including to the private sector. Continues in this position until 1979.
1978	All connections between Marwan & Mossad are severed. [Why?]
1979	Marwan, allegedly fired by Sadat, leaves Egyptian government service and the Board Chair of the AOI. By purchasing a 40% interest in Tradewinds cargo airline Marwan begins working with R. W. "Tiny" Rowland, famous British billionaire who held majority shares (60%) in Tradewinds through Lonrho, R's umbrella company since 1961 when recruited by former British intelligencer, Sir Joseph Ball. Rowland was already involved in arms dealings, particularly in Africa, and a bitter enemy of Mohamed al-Fayed. Also closely involved with such infamous international financial moghuls as Adnan Khashoggi, who'd introduced Marwan to the Muktoum brothers who'd then introduced him to Rowland.
	These connections draw Marwan deeper into the world of private international industrial finance, arms deals, and covert intelligence operations — thereby creating many new dangerous enemies, including al-Fayed. [BW: And providing ample room for the conspiracy theory of your choice.]
c.1980	M moves with wife Mona to Paris, living in his previously purchased luxury apartment at No.50 Avenue Foch.
	M becomes co-owner of Cyprus-based Overseas International Distributor (OID), which dealt in commercial aircraft, specifically as Boeing's exclusive distributor throughout most Middle Eastern countries. His partners were Fouad al- Zayat, a Syrian entrepreneur who'd founded the company in 1976; and Sheikh Kamal Adham, a Saudi who'd headed the Audi General Intelligence Directorate until 1977. Although OID closed in 1981, it revived as FN Aviation until 2003 when again changed names to FAZ Aviation, all still under al-Zayet.
1981, 6 Oct	Sadat assassinated by an Islamist cell in the military headed by a lieutenant who'd been inspired by a fatwa from Omar Abdel-Rahman, the "Blind Sheikh".
1981, winter	Marwan moves permanently to London, buying a large flat posh No. 45 Carlton House Terrace, overlooking St. James's Park in Mayfair. (By 2007 this 15-room suite with 7 bedrooms is valued at £4.4 million).

	Already quite wealthy, M advances toward (and allegedly into) the US$ billionaire level.
1984	M sells his £4 million House of Fraser shares to Rowland, which brings M into direct legal confrontation with the Al-Fayad brothers who are trying to take over the Fraser's department store empire, particularly its flagship, Harrod's.
	Al-Fayad accuses Rowland & Marwar of employing "dirty tricks" against him in this long struggle. The instrument is the Marwan-owned Octogon, a London-based security firm, which hired a special team of private detectives led by former senior police officer David Coughlin.
1985	The Al-Fayads finally succeed in buying (for £615 million) the House of Fraser, driving off Rowland (and Marwan) and exacerbating their mutual hatred.
1993	First public hints that Marwan was a double agent emerge in Gen. Zeira's book *Myth versus Reality* (in Hebrew), while citing only Richardson's book from the late 1980s. Israeli journalists bury this revelation as contrary to Israel's best interests. But "spiking" a news story doesn't mean it has stopped spreading by word-of-mouth — and spread it did.
1999, Sep	Israeli reporter Ronen "Roni" Bergman, then with *Haaretz*, alleges in articles in his paper that an otherwise unnamed "Bavel" ["Babylon"] was not only an Egyptian double agent but also cites the GARBO case (see above) as precedent, although he (or Bar-Joseph, "The Intelligence Chief who went Fishing in the Cold," p.237) fails to mention the key It's-Later-Than-You-Think ruse — limiting the revelation to the warning about Normandy. [On 21 Oct 2004 former *Haaretz* journalist Aluf Benn reported that he, having earlier figured out Marwan's identity, got his editor to assign the investigation of Marwan to Ronen Bergman who got confirmation from Israeli intelligence veterans, Marwan having declined an interview.]
2002, Sep	M again not explicitly named but readily identified and hinted to be Israel's "master spy" in London-based Israeli-British historian Ahron "Ronnie" Bregman's revised book, *Israel's Wars* (2nd edition, pp.74, 113-114). We now know that Bregman's source was Zeira.
2002, Dec	In a widely reported interview in London with Egyptian newpaper *Al-Ahram* Dr. Bregman again points toward Marwan.

	Marwan tells Egyptian press this is "an absurd detective story"; but Bregman, claiming the need to defend his reputation as an historian, now *explicitly names* Marwan as Egypt's double agent.
c.2002	Bregman interviews Marwan. Once, asked why the information he gave the Israelis was 4 hours late, M answers, "What difference does a few hours make?" REF: *Haaretz*, 28 Jun 2007. [BW: The difference did matter — it actually helped the Egyptians; although it did, perhaps inadvertently, hurt the Syrians, as I argue below.]
	On a subsequent interview, when Bregman asks what kind of book he was writing, M explained that "Everyone in Egypt, the whole system, worked to embarrass Israel." Bregman later tells *Haaretz* that "I concluded from this that he really was a double agent." REF: *Haaretz*, 28 Jun 2007. [BW: M's inclusive "everyone" does indeed read like a coy confession of his double agentry.]
2003, Sep	Howard Blum's book on the Yom Kippur War (*The Eve of Destruction*) hints (on the basis of his Apr 2002 interview with Marwan and his May 2002 interview with Zeira) that Marwan had been "The Source".
2003, Fall	A series of articles in the Yedioth Ahronoth by Ronen "Roni" Bergman & Gil Meltzer reveal fresh details of the depth of the Yom Kippur surprise, largely on the basis of recently declassified Israeli government documents.
2004	Zeira, in the 2nd edition of his book, also now *explicitly* names Marwan as Egypt's double agent. He even repeats Bergman's GARBO story as precedent, although I suspect Zeira himself had been Bergman's source for the GARBO connection.
2004, 31 May	Joel Beinen in article titled "The Good War" in American magazine *The Nation* names Marwan as "the In-Law" [The Source] on the basis of clues in books by Abraham Rabinowich & Howard Blum.
2004, Sep	Zamir, in an Israeli TV interview, denounces Zeira as a "traitor" for having been the source of Marwan's "outing" a genuine Israeli agent and calls for a trial for breach of security. Zeira sues for libel.
2004, 6 Oct	Despite these widely published charges & counterpunches, Egyptian President Mubarak is shown on Egyptian TV

	warmly greeting Marwan during a 31st anniversary Egyptian celebration of the Yom Kippur(Ramadan) War. This, some claim, is proof that Marwan had been Egypt's double agent.
2007, 17 Jan	Marwan reappointed a Director of the British Egyptian Society after long absence due to ill health — he'd reportedly had 3 heart surgeries. [BW: Evidence of improved health?]
2007, 14 Jun	Publication of the decision (made secretly on 25 Mar) of the Israeli court arbitrator, retired Supreme Court Vice President Theodor Or, dismissing Zeira's libel charge, deciding Zeira had indeed leaked Marwan's name to the media, and ordering that Zeira pay Zamir's legal fees.
2007, Jun	M tells American journalist Harold Blum he'll reveal more about his role at a news program in the USA; but 2 days before the scheduled taping he tells Blum he won't speak publicly until he finishes the final chapter of his book about the war. Says his title will be *October, 1973 — What Took Place*. [But does this manuscript really exist?]
	When Blum asks him, "Are you afraid?", Marwan replies, "Why would I be afraid? I was a soldier." Given Marwan's earlier service in the Egyptian Army, this phrasing is ambiguous as to admitting or denying double agentry. However, I read it as tending toward an admission — most patriotic spies would think of themselves as "soldiers", few spies motivated by money (as M told Mossad) would do so.
2007, 26 Jun	M agrees to meet Ronen Bregman next afternoon in London for an interview that would not happen.
	He'd also scheduled a meeting at London's British Egyptian Society to decide the amount of money to contribute it, according to the Society's treasurer, as reported in the Cairo *Al-Masry Al-Youm*.
2007, 27 Jun AM	
	That morning, during a video chat with his wife (who was visiting in Lebanon), Marwan tells her (or so she later said) that he was preparing to go to "New York" [!] later that day.
2007, 27 Jun PM	
	Marwan, age 63 [sic], dies almost immediately after fall from 4th floor (British style, 5th floor in American style) balcony of his posh London suite in Carlton House Terrace.

Is it coincidence that his death came only 13 days after publication of of the Israeli arbitrator's naming Marwan? Most commentators assume cause-and-effect, although I don't understand this thinking. After all, as seen above, Marwan had already been "outed" during the decade 1993-2003 among so many journalists that I assume this secret had become known to everyone with any interest in the matter.

At the instant Marwan began his fall (Cairo daily *Al-Masry Al-Youm* reported) 4 members of the board of directors of Marwan's chemical company, Ubichem, were in a hired room on the 3rd floor of the Institute of Directors Building at 116 Pall Mall. This room was only 15 yards across the courtyard and 2 floors below M's suite. They were waiting to meet that afternoon with M. These men were Mike Parkhurst (M's British secretary), Essam Shawki (acting chairman & 10% co-owner of of M's Ubichem and husband of Azza, a long-time Marwan secretary & daughter of Sadat's former secretary, Fawzy Abdel Hafez), and two otherwise unidentified "Hungarian men".

We now know that one of the Hungarians was Jozsef Repasi, a chemist who in 1996 had joined Marwan's Ubichem as founder-manager of its phamaceutical R&D division in Budapest. [BW: Conspiracy theorists will appreciate that he was an amateur orchidologist.]

Mr. Repasi told a *Times* reporter in Budapest of having seen the body pass his window and then looking up to notice two men "wearing suits and of Mediterranean appearance" on [what he later presumed was M's balcony] look down and then go back inside the flat.[146]

Mr. Shawki claimed that when Marwan was late he phoned M, at which point M said he'd dress and come in a half hour. They next saw M come onto his balcony acting strangely. Shawki phoned again and was told "I am coming, I am coming"; but then, looking back into the flat, said "Listen Shawki, I am not coming, I am not coming." One witness assumed Marwan had seen one or more persons [BW: Repasi's two men?] in the room. Shawki explicitly told police he saw M jump. [BW: Witnesses often misremember by elaborating, filling in the gaps. Perhaps Shawki saw M's falling body, assumed he'd jumped, and later remembered that "I saw him jump!"].

---

146 These details seem most accurately stated in *The Times* (13 Oct 2007).

One of the other four added that Shawki had reached M's body in time to detect a pulse but that it had stopped before the ambulance arrived.

2007, 27 Jun 4 PM

Marwan's housekeeper tells police she was working in the kitchen [by other accounts, with the cook] and heard nothing unusual, assumed she was alone [the cook?] except for M working in his study, and only learned of his fall around 4 PM when someone knocked at the apartment door and told her of it.

Some surveillance cameras in M's building were later found to be out of order.

2007, Jun-Jul

M's widow, Mona, fires Shawki. She claims, "He is a liar who said to have seen my husband committing suicide." She also claims that Shawki & his wife Azza had embezzled "millions" from Ubichem. In reply, Shawki sues for defamation. [With what outcome?]

British police express disapproval of Mona's public accusations of murder with foreign involvement.

The general knowledge that Marwan had been in poor and declining health — with a nervous disorder of his feet that required walking with a cane — fueled speculation of possible suicide. However, his sister reported saying she'd found Marwan in "good spirits" within two hours of his death. And his wife & two friends said he was too good a Muslim to suicide. Moreover, both Bregman and the treasurer of the British Egyptian Society noted that he'd made appointments with them for the next day.

Alternatively, it was speculated that perhaps he'd lost his balance and fallen. However, his son argued that the layout of the balcony combined with M's weaken physical condition would have ruled out either an accidental or suicidal tumble. Police also quickly rule out accident because the balcony railing was too high (4.5 feet), intact, secure, and blocked by a flower pot and air conditioning unit.

The Times reports that Marwan's shoes removed from his body disappeared at the mortuary — shoes that might yield forensic clues (such as dirt from the flower pot) to whether or not he'd climbed up and jumped.

I–95

Then there's the question of M's alleged manuscript memoir. Police suspect it never existed — only a story M had told Blum. However, widow Mona is reported as telling the police that her husband's unfinished manuscript had disappeared that day from his bank's safe deposit box along with other papers she thought might be connected with his fear of being murdered, a fear he'd expressed to her on 3 occasions. REF: Gordon Thomas, *Secret Wars* (NY: St. Martin's, 2009), 159-160.

Additionally, someone (Mona?, Blum?) describes the manuscript as 3 volumes of some 200 pages each together with the tapes on which they were based — plausibly circumstantial details that imply genuine tapes & manuscript.

2007, 2 Jul	Marwan buried with high honors in Cairo. Next day, Egyptian President Mubarak, while returning from official business in Ghana, asserts that Marwan "carried out patriotic acts which it is not time yet to reveal." Mubarak also said that no one other than Sadat and a few senior military officers knew the Yom Kippur War timing.
2007, Nov	M's elder son Gamal succeeds his father as Ubichem chairman. That Gamal keeps Mr. Repasi as head of the pharmaceutical division seems to exclude R as a suspect.
2008, Jan	British Metropolitan Police ("Scotland Yard") bogged investigation into Marwan's death is transferred to its Special Crime Directorate (SCD).
2008, late	British police spokeswoman confirms that police now know the identities of the two persons that Repasi had earlier reported seeing on the balcony immediately after Marwan's fall. [So, who are they?]
2009, Sep	M's younger son Ahmed still claims his father had been murdered.

END OF MARWAN TIMELINE

■ ■ ■ ■

I have five major problems with the semi-official Israeli claim that Marwan was an Israeli agent:

First, and most significantly, the suspicious timing of his Yom Kippur warning — almost a reprise of the above GARBO case, which we know had been calculated to mask GARBO's double agentry. Most strategic warnings come in timely fashion (see my *Stratagem*, Chapter 5). Many warnings come well after the event. But only rarely does strategic warning come in those few hours or days' window when the secret is out but no effective response is possible — what Liddell Hart called "Surprise Effect" without actual surprise. Even Bar-Joseph sees the only advantage to Israel in that the warning may have bought them enough time to mount their decisive counter-attack on the Syrian Golan Heights. But that was an improvised tactical fluke of genius & technological surprise (the first battle ever won by bulldozers). Moreover, I'd expect Sadat to be selfish enough to willingly take a low-risk by playing a bit fast-and-loose at his Syrian ally's expense.

Second, but almost as important, is that Marwan not only sold the attack *time* but the war *strategy*. Bar-Joseph's analysis:

> "Marwan gave Zamir the real Egyptian war plan—a plan that had been kept secret even from those in Egypt who knew about the coming war and of which the Syrians had not been aware at all. The plan could have given the Israelis a major advantage when fighting started. The fact that they did not use it effectively has nothing to do with Marwan's bona fide and a lot to do with AMAN's [Zeira's] poor performance at the beginning of the war."

This sounds like very strong — even conclusive — evidence. But it bypasses the decisive fact that Sadat's plan — his strategic goal — wasn't some all-out invasion to wipe out Israel but only a narrowly limited goal of bruising Israel just enough to a) wipe the shame of Arab defeat in the 1968 Six-Day War and, more importantly, b) pressure Israel to negotiate away at least some symbolically significant part of her spoils in that war. At most, Sadat expected his surprise attack to recapture the east bank of the Suez canal and occupy a large chunk of the Sinai Peninsula and then bring an obviously disadvantaged Israel to the negotiating table. This only seems counterintuitive. Sadat should and plausibly did want the Israelis to know his plan, his goal, so that they would not escalate to a total war for survival — the war that some Israeli intelligence officers believed would go nuclear.[147]

---

147 Private information in Apr 1975 from two Israeli reserve field-grade intelligence officers who'd been in the GHQ bunker continuously until the fateful night of 15/16 Oct 1973. When ordered to sleep that night, they embraced, knowing that the Suez front had virtually exhausted its reserves and assumed there were only two options. Either an exhausted Egypt would grind to a halt or a desperate Israel would go nuclear. My friends awakened from their bunker cots

Rather like Osama ben Laden after the Twin Towers, Sadat was as surprised by his own success as was his enemy.

Third and even less important, all intelligence services are rightly suspicious of "walk-in" agents like Marwan precisely because they are self-selected — in effect recruiting us. Much safer when we spot & recruit them — unless they've been deliberately making themselves visible as attractive candidates ("coat-trailing"). Moreover, it is suggestive that, having once secured the confidence of his Mossad handler ("Dubi"), Marwan insisted that he henceforward deal only through "Dubi". Mossad did so, although it meant keeping "Dubi" from further promotion.

Fourth, ambiguous but suggestive, was M's claim to Blum that he'd acted as "a soldier", as I speculated above.

Fifth, least important and only marginally so, is that no one gives any solid motive for Marwan's treason other than those fat payments to an already fairly rich man on the cusp of being far wealthier through safe & conventional corruption. Money is often the main motive for espionage, but double agents are even more likely to demand it, as did Marwan.

Additionally, many other of Bar-Joseph's more plausible arguments can be challenged for reasonable doubt. Most importantly perhaps is his point that M's two false alarms of Dec 1972 and Apr 1973 had been given in good faith, as they'd been the genuine original dates. But what if Sadat had only used these alarms to test the Israeli mobilization capability and timing? Surely that would be worth lowering an agent's rating for accuracy. Remember that Marwan's credibility had depended as much on passing documents as well as his verbal assessments.

No one of these five sets of evidence is conclusive. Indeed, none alone rises above a 50/50 probability. However, taking them as independent variables and applying the Product Rule, they combine to push the likelihood upward. Having begun largely agreeing with Bar-Joseph, my final reassessment now leans heavily toward the opposite hypothesis, namely that Marwan had been a double agent for the Egyptians — 65/35 or perhaps even 75/25.

■ ■ ■ ■ ■

It's surprising then that, after more than three decades, three key elements of the central mystery remain:

---

to surprising news of a third option — General Sharon's decisive war-ending end-run thrust across the Suez canal.

- Why did Gen. Zeira reveal the name of this agent? Bar-Joseph's answer is a convoluted psychiatric-type profile of Zeira that is simultaneously libelous, wildly speculative, and simply assumes some questionable points of fact. I question Bar-Joseph's portrayal of Zeira. In general, it doesn't fit my personal estimate of Eliahu Zeira. I knew Z moderately well in 1974 when we were both at the RAND Corporation. At that time he readily, regretfully, but undefensively admitted to me that he'd misread the evidence.

  More importantly, in freely speculating on the motive behind Zeira's double-agent theory, Bar-Joseph (2008, p.244) argues that Z would prefer to believe he'd been fooled by "a highly sophisticated Egyptian deception plan" than that he'd simply misread the evidence. But that theory overlooks two facts: First, Z isn't nearly as defensive about his misreading than B presumes. After all, he'd already faced, accepted, & admitted that error by 1974 when we met. Second, Z was no ordinary intelligencer but was proud of being both a political-military *deception* planner & operator and an amateur *magician*. I believe the most embarrassing and shame-faced thing for any specialists in deception, magic, or the con such as Z or myself would be to accept, much less publicly admit, the possibility that we'd been conned by a con. For us, unlike the proponents of the Wohlstetter-Betts Signal/Noise Model, being deceived is no excuse, no Get-Out-of-Jail-Free-Card.

- The far more important question: Was Israel's single most influential source of warning, Ashraf Marwan, a genuine Mossad agent (as generally believed by Israeli intelligencers) or was he an Egyptian double agent (as insisted by most Egyptian sources and former Israeli military intelligence chief Eliahu Zeira)?

- Is Ashraf Marwan's death connected to the question of whether he was an Israeli agent or an Egyptian double agent or some other reason? What caused his death? I see 7 plausible possibilities, 7 competing hypotheses — presented here in my assessment of their rising order of plausibility:

H1.

Accident. Virtually impossible given his physical incapacity & the balcony architecture, as noted by both his immediate family, the police, and a press photograph of the balcony.

H2.

Ordered by someone for personal reasons (discarded lover, cuckolded spouse, etc.). Highly unlikely, given the absence to date of any evidence other than vague rumors of M's earlier womanizing and separations from his wife. As access to the building was through a concierge, thieves or unwelcome visitors would not have easy entry to M's apartment.

H3.

Suicide. Unlikely, given M's state of mind at that time as reported by both immediate family & associates. If he intended suicide some other means would seem more plausible than the physically difficult one of climbing over his balcony parapet.

H4.

Assassinated by Egyptian Intelligence (presumably the GIS). Unlikely, unless — and there's no evidence — they'd come to believe Marwan had been an Israeli agent.

H5.

Assassinated by Israeli Intelligence (presumably Mossad). Only if the Israelis had secretly decided that M had indeed been an Egyptian double-agent. However, the only secret foreign intelligence service I'm aware of that went out of its way to assassinate long-retired opponents more for vengeance that any operational "necessity" was Stalin's. The only plausible motive for the Israelis to kill him *at this late date* would be to suppress his manuscript memoirs and post-publication press publicity. For any in current generation of Israeli intelligencers or politicians to neglect urgent current priorities to either conceal much less avenge their predecessor's blunders would be unprofessional. Or is that hoping for too much rationality?

H6.

Murdered as an act of pure revenge by a few former Israeli intelligence officers acting without the knowledge, much less the approval, of higher Israeli authorities. Possible, fanciful, and only slightly more plausible than *H5*.

H7.

Murdered by current business competitors. Entirely possible, quite plausible, and highly probable (say, 70-75% chance). In my opinion this is the most likely explanation in spite of the lack of evidence beyond the circumstantial facts a) that M was known to been & still was engaged in many lucrative but covert arms and other shady commercial deals that created some very powerful enemies; and b) the odd story (convenient alibi?) given by Marwan's four business associates who claimed to have witnessed his behavior leading to his death-fall.

■ ■ ■ ■ ■

Are these 7 hypotheses solvable mysteries? Further Egyptian sources should enable us to resolve the agent versus double agent theories of **H4** and **H5** and hence at least simplify the murder vs suicide and whodunnit hypotheses. **H6** and **H7**, if either is correct, will, as usual in these matters, most likely be revealed within a very few years by one or more talkative conspirator. Finally, I'm content to leave the question of General Zeira's motives to Israeli courts, although it seems to me that when it comes to any findings of intelligence failures that there was plenty to share.

In sum, even long after the event, the Israeli intelligence services and high command have generally remained convinced that The Source (Marwan) had been their controlled agent and not an Egyptian double agent. On the contrary, I think it likely but not *highly* likely that, except for Zeira, the Israelis are wrong. I remain rather more than half persuaded (say 60/40) that Marwan was Egypt's double-agent. My residual doubts are due entirely to Bar-Joseph's analysis. It's the best to date — far more detailed and persuasive than the many others, including CBS-TV's very weak *60 Minutes* segment of 10 May 2009. But nowhere nearly as conclusive as Bar-Joseph asserts.

7. Double-Agent Systems: Dead or Alive?

No one doubts that double agents will always be with us — both theirs & ours. But only as *individuals*, not as part of an orchestrated *system*. Popular wisdom today assumes that double-cross agent systems are a dead issue. The great majority of present-day intelligence analysts state with nostalgic regret but unquestioned faith that the combination of circumstances that made a double cross *system* possible in WW2 was so unique that we shall never see its likes again. This is surely wrong — a pessimistic leap too far into an unknown future.

To the contrary, several plausible resuscitation scenarios seem not merely possible but likely. All they require is two minimal conditions: (a) complete control over any single channel of communication; and (b) confidence in at least one feedback channel from the enemy that the system is working unsuspected. It does seem unlikely that any deceiver will ever again take control of *all* enemy agents, as had the British XX Committee or the Germans in Holland in WW2. However, a deceiver today need only control a single isolated enemy network to work a double-cross system. Similarly, our deceiver may never again enjoy such a degree of welcome *feedback* from the enemy's wireless communications through secret code-reading access. But today's deceiver only need have this feedback from any one of many possible channels — for example a HUMINT channel.

So, let's not dismiss the double-cross scenario as passé. While unlikely we'll again see a double-agent system run on anything like the vast scale achieved by Clarke & Masterman, I predict several smaller-scale ones. Indeed, given the weak points — those human & technological vulnerabilities that emerge from time to time in all intelligence services — it's likely we'll see a few complete takeovers of at least one significant component of an opponent's intelligence service.

Let's hope such attractive intelligence & security vulnerabilities won't become a fixed part of *our* system. However, unfortunately our track record of self-deception in underplaying our own vulnerabilities to enemy deception suggests that this is an urgent matter. The simplest solution is to seed a few deception analysts throughout the more vulnerable units in our intelligence and security organizations and then, ideally, coordinate them through some counterdeception group.

8. Limits of the Game of Turnabout: Simple Cycles or Infinite Layers?

> GENERAL: Incidentally, they know your code.
>
> AMERICAN AMBASSADOR (beaming): We know they know our code … We only give them things we want them to know.
>
> [next scene]
>
> GENERAL: Incidentally, they know you know their code.
>
> SOVIET AMBASSADOR (smiling): We have known for some time that they knew we knew their code. We have acted accordingly — by pretending to be duped.
>
> [next scene]
>
> GENERAL: Incidentally, you know — they know you know they know you know.
>
> AMERICAN AMBASSADOR: (genuinely alarmed): What? Are you sure?
>
> — Peter Ustinov, *Romanoff and Juliet* (1957), Act 2

One example of complexity often quoted by intelligencers is the old you-know-that-I-know-that-you know bit as nicely parodied by Peter Ustinov in his delightful play, *Romanoff and Juliet*. It's a Cold War comedy with a serious bite. Therein the General, the cunning head of a tiny democracy, ends three successive meetings with the diplomatic reps of the two rival Great Powers with the famous lines that open this chapter.[148]

Not surprisingly, Ustinov's charming story became a favorite among Cold War intelligencers. The CIA's James McCargar, writing as Christopher Felix, was one of the first to recognize that it reflected the essence of counter-espionage (CE) operations to "hilarious perfection."[149] And RAND's William R. Harris

---

[148] Peter Ustinov, *Romanoff and Juliet* (New York: Warner Chappell, 1957), Act 2. This play premiered in 1956 in London and Ustinov directed and acted in the 1961 movie version.

[149] Christopher Felix [pen name of James G. McCargar], *A Short Course in the Secret War* (New York: Dutton, 1963), 143-144.

reminded me of its relevance to CE in 1969 when I made it a hinge theme in *Codeword BARBAROSSA* (p.151).

Question: Just because this you-know-that-we-know-that-you-know kind of infinite regression can be presented in fiction, does this mean that it can work in real-world situations?[150] Not, I suggest, beyond the first exchange between 2 persons. Consequently this model's cycle logically has only 4 pairs of knowledge conditions:

>CONDITION 1: A knows the situation and B doesn't.

>CONDITION 2: A learns, so that now both A & B know the situation.

>CONDITION 3: B knows the situation and A doesn't.

>CONDITION 4: B learns, so that finally both know the situation.

>CONDITION 5 = CONDITION 1, that is, the cycle begins to repeat as soon as A learns this new situation.

I further suggest a close analog in the classic Agent/Double Agent Problem where with only 1 person being recruited there are logically only 2 conditions:

>CONDITION 1: A has been recruited by X to be his Agent, namely to coin a term, Singled.

>CONDITION 2: A has been turned by Y to be his Agent, that is, Doubled.

>CONDITION 3: = CONDITION 1, that is, the cycle begins to repeat so that Single = Triple and Double = Quadruple.

To put this less formally, at first sight this seems to be what might be called the Paradox of the Multiples. Here, as in the single/double/triple/quadruple agent puzzle, we have another endless chain of twists and turns. But, again, it is one that Occam's Razor dices into neat pairs. This simpler model has only two conditions: I-know-and-you-know-I-know versus I-know-and-you-don't-know-I-know. The next step up, I-know-that-you-know-that-I-know, can be logically reduced to the first condition, turning the open-ended chain of logic into a tight cycle. Some persons will, of course, prefer to view it as a never-ending chain; but why make life unnecessarily complicated.

Bill Harris was also fond of quoting these three passages from Ustinov's comic play and movie. He accepted its multilayered model. However, he also

---

150 An even less realistic problem of infinite *futures* was posed by Borges (1942) in his famous detective story, "The Garden of Forking Paths."

recognized that each additional layer adds a proportionate risk of failure. Thus he wrote that:[151]

> Chess players—of whom the Russians are among the best—will appreciate the endless layers that can be added: counter-counter-measures, and beyond. But the sheer complexity and bureaucratic inefficiency should cause most of these additional layers to collapse of their own weight.

Which is the more useful model for picturing the sequence of cross, double-cross, triple-cross, etc. Infinite regression or cycle? Is it an infinite regression where the sequences of I-know-that-you-know or EM-to-ECM-to-ECCM, etcetera never end? Or is it a cycle where the sequence turns back on itself so that the triple agent equals a simple agent and the quadruple equals a double? Although I've discussed this problem elsewhere,[152] it's necessary here to elaborate & apply it to this specific context of the turnabout game.

Both models are useful; both have their place. In the cycle model Occam's Razor keeps it simple; in the infinite regression model Crabtree's Bludgeon magnifies certain details. When we focus on the *sequence* of actions and reactions, the Bludgeon model of infinite regression gives a closer analogy. For example, although our hypothetical "triple" agent can be recycled as a simple agent, that model surely oversimplifies the psychological effect of tripling on most individuals. In other words, most persons can double-cross and keep track of their ultimate loyalty with the only price for some being, at most, twinges of guilt. By contrast, most persons who triple-cross risk cognitive consequences of dysfunctional confusion or even catastrophic panic. Having said this, I find the simpler cycle model handier for most purposes.

Note that these opposing models of regression vs cycles of *mutual knowledge* don't involve the very different question of the levels or layers of *sophistication* in planning or operation that deceptions can take. That is the notion that deceptive ability comes in "degrees" (levels or layers) and that the party with the higher "Degree of Cunning" will usually outwit the inferior.[153]

---

151 W. R. Harris (1972), Section III ("Counter-Deception Planning: Methodologies").
152 Whaley, *The Maverick Detective*; or, *The Whole Art of Detection* (in progress).
153 The earliest explicit fictitious military version of this notion appears in Swinton (1908).

Appendix:
The Double-Crosser's Dictionary

> In the nomenclature struggle, who names an issue usually carries the day.
>
> — William Safire, "Language" column, *International Herald Tribune*, 15 Aug 2005, p.9

blowback

n. The unintended consequence of a failed operation rebounding upon the planner. Coined in this sense in 1954 by a CIA report (recently declassified) on the Agency's successful coup d'état that overthrew Iran's Mossadeq regime only to have it replaced by an even less desired regime. REF: *Wikipedia*, "Blowback (intelligence)" (accessed 20 Nov 2009).

CIT: *Clandestine Service History—Overthrow of Premier Mossadeq of Iran—November 1952–August 1953* (CIA: 1954).

Interestingly, this term didn't reach standard dictionary status until 1973. REF: Merriam-Webster's *Collegiate Dictionary* (11th Edition). The original meaning as "blow back" & "blow-back" has been traced back to referring to the blowing back of an engine. REF: *OED2*.

bluff

v. To attempt to deter or mislead by a pretense of strength, confidence, or assurance. Coined in America by 1839 and soon applied to the game of poker, which was originally called "bluff". Compare **double-bluff**.

REF: Leighter, Vol.1 (1994), article "bluff".

> Of the two possible motives for Bluffing, the first is the desire to give a (false) impression of strength in (real) weakness; the second is the desire to give a (false) impression of weakness in (real) strength. Both are instances of inverted signaling — i.e. of misleading the opponent.
>
> — Neumann & Morgenstern, *Theory of Games and Economic Behavior* (1944; 2nd ed 1947), 189

n. The act of making a bluff. Coined in America by 1845.

bluffing

   adv. Attempting a bluff. Coined by 1850.

       REF: *OED* (2nd Edition).

break camouflage

   v. To reveal a disguise, to unmask. Camoufleur's jargon by 1942.

camouflage

   n. The hiding or simulating of physical objects, as with paint, foliage, netting, dummies, etc. Originally and particularly applied to military equipment or personnel. Coined in 1915 (by 12 February) in French and diffused into the English language later that same year.

       CIT: 31 Dec 1915, Douglas Haig, letter, as "camouflage work"

       The important distinction between **negative camouflage** (hiding) and **positive camouflage** (showing) was coined as early as 1918, their synonyms of **passive** and **active** camouflage. The distinction between **strategic camouflage** and **tactical camouflage** was coined in 1920. A person who engages in camouflage is called a **camoufleur**.

   v. To hide by disguise. Standard English by 1917. Opposite of to **break camouflage**.

charc

   n. Abbreviation of **characteristic**. Coined in 1979 by J. Bowyer Bell in our draft "Dublin Papers"; first published in Bowyer (1982), 61.

counter-espionage, counteresponage (CE)

   n. The penetration of an opponent's intelligence service. Counter-espionage, usually known by its acronym CE, is a widely misunderstood branch of secret operations, particularly in the U.S. Intelligence Community, where it is often confused it with counter-intelligence (CI). CI may involve agents and sensors that passively listen in on an opponent's secrets; CE may also involve agents who penetrate an opponent's intelligence services in order to influence or sabotage their policies and programs.

       REF: Felix (1963), 143-152.

       CIT: Masterman (1945/1972), 1, as "deception and counterespionage".

       This term was in use in France as *contr'espion* as early as 1793 (by Dlandon).

counter-stratagem

    n.   Equals both the detection of deception and, by extention, **turnabout** or **double-cross**. Coined in 1968 by William R. Harris, adopted by Whaley (early 1970s) and Scot Boorman (1972), and independently suggested by Jeff Busby (on 27 Mar 1984). I don't recommend using this unappealing & ambiguous term. Fortunately, it didn't catch on, leaving it as what lexicographers dismiss as a "nonce word".

cross

    n.   A swindle, cheat, or betrayal such as (originally) in a fixed prizefight. Often in the phrase **on the cross**. Opposite of **square**, **on the square**. Compare **double-cross**.

        CIT: 1802 in Partridge, *Dictionary of the Underworld*, as quoted in Lighter.

    v.   To swindle, cheat, or betray.

        CIT: 1821 *Real Life in London*, as quoted in Lighter.

cut-out

    n. An **agent** who is acting as an intermediary between another **agent** and the latter's **case officer**. Usually designed to assure anonymity. Compare a **drop**.

    REF: Felix (1963), 59.

double agent

    n.   An enemy agent whom you have coopted ("turned") to work back at them for you, a "double-cross agent".

        Special Means was the **cover name** in WW2 among British & American deception planners. Popular writers and the media regularly mislabel simple **penetration agents** as double agents.

        CIT: Masterman (1945/1972), 1, as "The use of double agents in time of war is a time-honoured method both of deception and counterespionage."

        CIT: J. H. Bevan, letter to Bedell Smith, 11 Dec 1944, urging non-publication of Ingersoll's official history of FORTITUDE. REF: Holt (2004), 666.

        CIT: Begoum (1962).

        CIT: Goldman (2006), 45.

        REF: Felix (1963), 146.

a. Adjectivally as **double agent system**, etc.

CIT: Masterman (1945/1972), 3, as "double agent system"; 8, as "a double agent case"; 16, as "double agent information".

A triple bluff is an action intended to be perceived as a double bluff, but which is in fact merely a bluff. A quadruple bluff is the next in sequence after a triple bluff. REF: *Wiktionary*, "double bluff", "triple bluff", "quadruple bluff" (accessed 21 Sep 2009).

double bluff

n. An action intended to be perceived as a bluff that is not.

v. To trick someone by appearing to bluff, while not bluffing.

double-cross

n. An act of double duplicity such as (originally) when a prizefighter or jockey who has agreed to throw a fight or race breaks his word.

CIT: 1826 English Spy, as quoted in Lighter.

v. To **double-cross**. Compare **cross**.

CIT: 1901 Lighter.

a. Adjectivally as in a "double-cross agent".

CIT: Masterman (1945/1972), 17, as "double-cross work"; 21, as "each double-cross case".

drop

n. A person or place where written communications can be left for transmission without personal contact between transmitter & receiver. Usually designed to assure anonymity. Also called a **dead drop**. Cut also **cur-out**.

REF: Felix (1963), 59.

Electronic Counter Counter Measures (ECCM)

n. "Electronic counter-countermeasures (ECCM) is a part of electronic warfare which includes a variety of practices which attempt to reduce or eliminate the effect of electronic countermeasures (ECM) on electronic sensors aboard vehicles, ships and aircraft and weapons such as missiles. ECCM is also known as electronic protective measures (EPM), chiefly in Europe. In practice, EPM often means resistance to jamming." REF: *Wikipedia* definition (accessed 29 Nov 1990).

Electronic Countermeasures (ECM)

    n.   "Electronic countermeasures (ECM) are a subsection of electronic warfare which includes any sort of electrical or electronic device designed to trick or deceive radar, sonar, or other detection systems like IR (infrared) and Laser. It may be used both offensively or defensively in any method to deny targeting information to an enemy. The system may make many separate targets appear to the enemy, or make the real target appear to disappear or move about randomly. It is used effectively to protect aircraft from guided missiles. Most air forces use ECM to protect their aircraft from attack. That is also true for military ships and recently on some advanced tanks to fool laser/IR guided missiles. Frequently is coupled with stealth advances so that the ECM system has an easier job. Offensive ECM often takes the form of jamming. Defensive ECM includes using blip enhancement and jamming of missile terminal homers." REF: *Wikipedia* definition (accessed 29 Nov 1990).

notional

    a.   False, as in a "notional order of battle". This fine old word was recoined in its technical military and counter-espionage sense around 1941 by Brigadier Dudley Clarke, then Chief of "A" Force, the British deception planning team in Cairo.

one-ahead

    a.   Said of the method of keeping a step ahead of the spectators while the deception unfolds; a specific type of **anticipation**. A one-ahead mentalist card trick was used and exposed by magicians as early as 1693. Subsequently applied to billet-reading by spiritualistic mediums in the mid-1800s. Magicians jargon by 1930.

out

    n.   Any method introduced to salvage an effect if the originally method fails. Magicians jargon by 1938.

playback

    n.   To turn an opponent's deception around on her or him.

        CIT: Felix (1963), 15i, as "playing it back".

        CIT: Mangold (1991), 40, "The trade calls this 'playback.'"

red flag

n. To knowingly & ostentatiously display through specific words of actions a signal associated with classic patterns of a deception operation, ones that the intended victim should perceive as obvious but usually doesn't.

CIT: Coined in this deception-related sense in Stech & Elssaesser (2003).

CIT: Housworth (2005 & 2008).

sucker

a. Said of any effect where the magician leads the spectators to believe they have detected the method and then works a double bluff to surprise them even more.

Originally American slang since 1838 meaning a greenhorn, someone figuratively still sucking at the teat. REF: *OED/2*.

Thence magicians jargon by 1905 as a **sucker trick**; 1924 as **sucker effect**; 1924 as **sucker gag**.

sucker effect

n. A double bluff; the method of leading the audience to believe they have detected a trick's method and then pulling a double bluff to surprise them even more; a **sucker gag**; a **sucker trick**.

The sucker effect is the magicians' equivalent of the gambling sharps' "The Cross" (by 1802). The more usual sucker tricks include Acrobatic Silks, Chinese Wands, Die Box, Hippity Hop Rabbit, Fraidy Cat Rabbitts, Kling-Klang, Monkey Bar, Rattle Bars, Soft Soap, Stung and Stung Again, Sucker Card Box, and Sucker Silk. Large-scale stage illusions that incorporate sucker effects are Backstage, Bangkok Bungalow, and the Duck Vanish. Otherwise standard tricks that can be given a sucker twist include the Color Changing Shoelaces and the Egg Bag. The dramatic conclusion of all sucker tricks is called the **blowoff**.

Magician's jargon by 1924. Also called a "sucker routine", "sucker climax", "sucker finish", and "sucker ending". Or simply a "hoax" or "fake expose".

sucker gag

n. A **sucker effect** being the slightly more usual synonym.

Magician's jargon by 1924.

sucker move

n. A **feint** (see).

Magician's jargon by 1924.

Bibliography

This bibliography gives the sources on turnabout most frequently cited in the text. All other sources are fully cited at footnotes.

As a bonus, 5 items are added that give additional cases of turnabout. These are Bayman (2007), Begoum (1962), Shaffer (1970), Shultz (1999), and R. Weiss (2007).

NOTE: "LOC" indicates convenient locations of copies — either on the Internet or, when marked "BW", in the Whaley Collection at the CIA Library.

Bar-Joseph, Uri
    "The Intelligence Chief who went Fishing in the Cold War: How Maj. Gen. (res.) Eli Zeira Exposed the Identity of Israel's Best Source Ever," *Intelligence and National Security*, Volume. 23, No.2 (Apr 2008), 226-248.

    LOC: Internet (JSTOR).

Barbier, Mary Kathryn (ca.1957-)
    D-Day Deception: Operation Fortitude and the Normandy Invasion. Westport: Praeger, 2007.

    LOC: BW.

Barkas, Geoffrey (1896-1979)
    The Camouflage Story. London: Cassell, 1952.

    LOC: BW.

Bayman, Anna (1976-)
    "Rogues, Conycatching and the Scribbling Crew," *History Workshop Journal*, Vol.63, No.1 (2007), 1-17.

    LOC: BW (copy); Internet.

        On the state of con artistry from the mid 1500s through the early 1600s. Includes a few instances of double-crossing (pp.8, 15n33).

Begoum, F. M. [pseudonym of John P. Dimmer, Jr.] (ca.1922-2005)
    "Observations on the Double Agent," *Studies in Intelligence*, Vol.6, No.1 (Winter 1962), 57-72. SECRET [declassified 1994].

    LOC: Internet (CIA).

The best open-source field manual on the care and feeding of double agents.

Dimmer ("Begoum"), a 1942 University of Maine graduate in civil engineering, had been with the CIA for two decades when retired in 1973 as Chief of Station, Bonn, West Germany.

Book of Daniel
Manuscripts ca.150 BC-ca.1 BC.

R. H. Charles (*editor*), *The Apocrypha and Pseudepigrapha of the Old Testament in English,* Vol.I (Oxford: The Clarendon Press, 1913), 638--664.

The *Book of Daniel* is widely considered one of the books of the Apocrypha. Although conventionally dated around 570 BC, most modern scholars lean toward a later date of around 150 BC, with three chapters of likely even more recent origin. Two of those chapters are brief but important early tales of detection, both specifically involving abductive inference and both being examples of turnabout. And both were evidently later inserted in the *Book of Daniel* (together with one other unrelated chapter) in short segments now known as "Additions to Daniel". Both feature the Jewish prophet Daniel as the detective, the deception analyst:

"Bel and the Dragon" story (by 100 BC) in which Daniel forms an abductive hypothesis and then tests it by setting a cunning tripwire type of trap to expose the deceptions of the priests of Bel (Baal). For this story see pp.652-664 in the above edition and also Bruce Cassiday (*editor*), *Roots of Detection: The Art of Deduction before Sherlock Holmes.* New York: Frederick Ungar, 1983), 12-14; and Nickell (1989).

"Susanna and the Elders" story (by 1 BC) in which Daniel resolves a classic He Said/She Said dilemma by clever separate interrogations of the three suspects that reveal the fatal discrepancy between the stories of the two Elders, thereby proving their false testimony. For this story see pp.638-651 in the above edition.

Connell, Richard (1893-1949)
"The Most Dangerous Game," *Collier's Weekly* (19 Jan 1924).
LOC: Internet.

A classic short story illustrating the Double Back Principle.

Crowdy, Terry

Deceiving Hitler: Double Cross and Deception in World War II. Oxford: Osprey, 2008.

LOC: BW.

> The most detailed & comprehensive account of the British double agent system. Followed closely by Rankin (2008) and Masterman (1972).
>
> Crowdy is an ex-rock musician turned military history buff. He has been publishing books at a more than one-a-year rate.

Delmer, Sefton (1904-1979)

The Counterfeit Spy. New York: Harper & Row Publishers, 1971.

LOC: BW.

> Biography of "GARBO" one of the top double agents for the British Double-Cross Committee in WW2. This was one of the earliest public accounts of British double-cross planning and operations. The author had good access to intelligence sources. His main source for this book was the personal recollections of Col. Noël Wild and the then secret documents provided by him, including a copy of Roger Hesketh's official top secret history of FORTITUDE. The fact that Wild's disclosures were unauthorized led to the official suppression of Delmer's British first edition. See also Pujol & West (1999) and Harris (2000).

Foot, M. R. D. (1919-)

SOE in France: An Account of the Work of the British Special Operations Executive in France, 1940-1944. Revised edition, London: Cass, 2004.

LOC: BW.

Garcia, Eric (1972-)

Matchstick Men: A Novel. New York: Villard, 2002.

> Dark novel by an American mystery novelist about a con game that gives *The Sting* (1973) one more twist of the tale by pitting the con artist protagonist against another con artist. It also inserts into the large cast of characters an even more savvy version of *Paper Moon's* Addie Pray. The novel is appropriately cynical, its Ridley Scott/Nicholas Cage movie clone, also titled *Matchstick Men* (2003), is Hollywood syrup.
>
> I include this work here only because the movie version was used as a case study in the deception/counterdeception hypergame

analysis in Mark E. Mateski, *Reciprocal net assessment: Introduction* (Version 1.0, © Alternative Analysis, LLC, 2009), 76-82.

Giskes, H. J. (1896-1977)
London Calling North Pole. London: William Kimber, 1953.

LOC: BW.

Hall, R[obert] Cargill (ca.1937-)
"Postwar Strategic Reconnaissance and the Genesis of CORONA," in Day, Dwayne A. Day, John M. Logsdon, and Brian Latell (*editors*), *Eye in the Sky: The Story of the Corona Spy Satellites* (Washington, DC: Smithsonian Institution Press, 1998), 86-118.

> On the American CORONA satellite program. Includes the scandalous use of Amrom Katz & Merton Davis for deception purposes (pp.113-114, 272n83-84, 273n85).

> Hall was a prominent American aviation historian, formerly with NASA, then with the USAF, and finally Historian Emeritus with the NRO.

Harris, Tomás (1908-1964)
GARBO: The Spy Who Saved D-Day. Richmond: Public Record Office, 2000. Introduction by Mark Seaman.

LOC: BW.

> This is the original official post-action report on British-run double-cross agent GARBO (Juan Pujol), as written & edited by Tomás Harris, GARBO's case officer during WW2 with M.I.5's section B1a. Harris's original classified report titled "Summary of the Garbo Case 1941-1945" was released by MI-5 in January 1999. See also Delmer (1971), where Harris is disguised as "Carlos Reid" and Pujol as "Antonio Jorge"; and Pujol & West (1985).

> The lengthy Introduction (pp.1-30) by Seaman is surprisingly weak. It completely misses the main point that Masterman's Double-Cross Committee was running a complex and interlocking *system*. Seaman thinks it was only more agents operating in isolation than had been previously worked by intelligence services.

Harris, William R. (1941-)
Counter-Deception Planning (Draft Manuscript). [Santa Monica, CA:] RAND Corporation, 13 Jul 1973, 169 unnumbered pages. Labeled as RAND Report R-1230-ARPA.

LOC: BW (copy from Rocca Library); Raymond G. Rocca Library on Soviet/Russian Intelligence, Security & Deception; scarce.

> This rough draft manuscript is a substantially expanded version of the author's previously unclassified papers. Its findings were reported in Apr 1972 to an ARPA-sponsored conference on strategic planning. Dr. Harris told me that this paper never moved beyond draft status.

Heuer, Richards J., Jr. (1927-)
"Nosenko: Five Paths to Judgment," *Studies in Intelligence*, Vol.31, No.3 (Fall 1987), 71-101. Originally classified SECRET.

LOC: BW; Internet.

> Closely reasoned analysis of the Nosenko-Angleton controversy that was waged within the CIA since the apparent defection of Soviet KGB intelligence officer Nosenko to the Agency in 1968. Was he a genuine defector or a Soviet plant as Angleton argued? Case studies like this have a certain stand-alone interest but they can contribute to theory development only if taken as models suitable for replication and creation of a database of similar cases. Like Wohlstetter's superb but interestingly flawed *Pearl Harbor* (1962) and Whaley's *Codeword BARBAROSSA* (1973), this fine paper by Heuer was so intended. Heuer himself later (1998) explained to J. Ransom Clark that:
>
>> "The long-term value of this article is not what it says about Nosenko or Angleton, but the lessons about how bona fides analysis in general should be done." (Comment to J. Ransom Clark, Apr 1998).
>
> Strongly endorsed by both J. Ransom Clark and Frank Stech and seconded by myself. See also Winks (1987/1996), Edward Jay Epstein, *Deception: The Invisible War Between the KGB and the CIA* (1989); Tennent H. Bagley, *Spy Wars* (2007), and Holzman (2008).

Holzman, Michael (1946-)
James Jesus Angleton, the CIA, and the Craft of Counterintelligence. Amherst: University of Massachusetts Press, 2008.

> A much overrated biography of the CIA's controversial chief of counterintelligence from 1954 until his resignation in 1974.

Dr. Michael Howard Holzman (1976 UC-San Diego PhD in Literature) is an American freelance writer & consultant on education.

Housworth, Gordon (1945-)
"Signals, Sprignals and Noise: Deception Analysis in Financial Events," *The Journal: A publication of Stout/Risius/Ross* (Fall 2005), 20-22.

LOC: BW; Internet.

Applies the Whaley Incongruity Analysis model (plus inputs from W. R. Harris, F. Stech, and J. Boyd) to the 2001 Enron Corp collapse. The author recommends this model for identifying the sprignals (spurious signals) that red-flag an on-going deception in order to predict likely future events in the corporate environment. Houseworth refers to the Incongruity Analysis theory as "this current deception paradigm". This is, I believe, the second published effort to apply Incongruity Analysis theory to events in the world of business.

Gordon C. Housworth, Managing Partner at Intellectual Capital Group LLC. (ICG), has been a long-time correspondent with MITRE's Frank Stech.

Jones, R. V. (1911-1997)
The Wizard War: British Scientific Intelligence, 1939-1945. New York: Coward, McCann & Geoghegan, 1978.

LOC: BW.

Reflections on Intelligence. London: Heinemann, 1989.

LOC: BW.

Kahn, David (1930-)
The Codebreakers: The Story of Secret Writing. New York: Macmillan, 1967.

LOC: BW (revised edition).

Lippman, Laura (1959-)
What the Dead Know. New York: William Morrow, 2007, 376pp.

LOC: BW.

Mystery novel about an impostor, the mystery being who and why. The impostor's main problem, "the tricky part" (p.42) is "Not knowing what she should know but remembering what she *wouldn't* know." This is an astute idea that I don't find elsewhere although it is the one essential incongruity that proves imposture.

Lippman, an award-winning American detective novelist since 1997, had been a newspaper reporter for 20 years until 2001.

Mangold, Tom (1934-)
Cold Warrior: James Jesus Angleton: The CIA's Master Spy Hunter. London: Simon & Schuster, 1991.

> A competent early biography. Concludes that Nosenko was a genuine defector and not a Soviet penetration agent. The contrary view is vehemently held by Bagley (2007) who completely discounted (p.219) Mangold: "His book accurately reflected CIA's defense of Nosenko and was thus studded with error, omission, misrepresentation, and invention, and colored by emotional bias for Nosenko and against his detractors." See also Winks (1987/1996), 322-438; Heuer (1987); Bagley (2007), and Holzman (2008).

Marks, Leo (1920-2001)
Between Silk and Cyanide: A Codemaker's War, 1941-1945. London: HarperCollins, 1998; New York: The Free Press, 1999.

LOC: BW.

> Perceptive and delightful memoirs of SOE's chief codemaker in WW2. Marks was much more than a cryptographer and cryptanalyst; he was a brilliant intelligence analyst. He comes closer than any others I know of in his ability to explain to outsiders how researcher-analysts go about their work — the grungy digging for facts, the timely value of hunches, the wrestling with hypotheses, the excitement of the chase, the elation of discovery, and the frequent frustrations of politics and bureaucracy. See also Giskes (1953).

Leopold Samuel Marks was the London-born son of the proprietor of London's fine antiquarian bookshop, Marks & Company, which was made world-famous in reader Helene Hanff's 1970 best-selling memoir, *84, Charing Cross Road*.

Masterman, J. C. (1891-1977)
The Double-Cross System in the War of 1939 to 1945. New Haven: Yale University Press, 1972.

LOC: BW.

An insider's view. Written at the end of the war as the official classified history of that committee. Published here in slightly abridged form. An important corrective to Masterman's occasional exaggerations is Mure (1980), although Mure sometimes overplays his criticism as pointed out by John Campbell in the below review. See also Wheatley (1980).

During WW2 John Cecil Masterman, an Oxford don and mystery story writer, was a British M.I.5 official and Chairman of the Twenty Committee, so-called from 20 = XX = double-cross. Knighted in 1959.

On the Chariot Wheel: An Autobiography. Oxford: Oxford University Press, 1975.

LOC: BW.

Includes Masterman's reflections on his work with the Double Cross Committee (pp.211-226) and his calculated skull-duggery in getting his history of that double-agent organization published (pp.348-367).

Moss, Norman (1928-)
The Pleasures of Deception. London: Chatto & Windus, 1977.

LOC: BW.

A fine collection of selected case studies of hoaxes and hoaxers, covering the broad array from con men to soldiers. Much enhanced by the journalist author's perceptive analyses of the psychology of deceivers and deceivees.

Mure, David (1912-1986)
Practise to Deceive. London: William Kimber, 1977.

LOC: BW.

Also in PB as *The Phantom Armies* (London: Sphere, 1979, 316pp).

A memoir-history of "A" Force, the British military deception planning team in Cairo during WW2. Major Mure had served on its staff 1942-44. Some serious errors due mainly to the author's lack of access to the archives.

Master of Deception: Tangled Webs in London and the Middle East. London: Kimber, 1980.

On British Army Colonel (later Brigadier) Dudley Clarke.

Naftali, Timothy J. (1962-)
"ARTIFICE: James Angleton and X-2 Operations in Italy," in George C. Chalou (*editor*), *The Secrets War: The Office of Strategic Services in World War II* (Washington, DC: National Archives and Records Administration, 2002), 218-245.

LOC: Internet.

A near-definitive account.

Dr. Timothy James "Tim" Naftali (1993 Harvard PhD) is a former Associate Professor of History at U of Virginia.

"Ole Luk-Oie" [penname of Maj-Gen. Sir Ernest Swinton] (1868-1951)
"The Second Degree," *Blackwood's Magazine*, Vol.183, No.1109 (Mar 1908), 317-335.

LOC: BW (copy).

Reprinted in Whaley, *Readings in Political-Military Counterdeception* (FDDC,2007), Chapter 3.6.5.

Swinton's story makes, I believe, the first explicit statements (Poe's having been somewhat unclear) that deceptive ability comes in "degrees" (levels or layers) and the party with the higher "Degree of Cunning" will usually outwit the inferior (Chapter 8). Moreover, it also illustrates the old "Know-your-enemy" M.O. category (Sections 5.1 & 5.2).

And it's the earliest fictional tale of military counterdeception; although I didn't realize this back in 1977 when I wrote the second, "The Door to Death."

Pujol, Juan (1912-1988); with Nigel West [penname of Rupert Allason] (1951-)
Garbo. London: Weidenfeld and Nicolson, 1985.

LOC: BW.

Rankin, Nicholas (1950-)
Churchill's Wizards: The British Genius for Deception 1914-1945. London: faber & faber, 2008.

LOC: BW.

A comprehensive survey. "Nick" Rankin, a British writer and broadcaster, was then the BBC World Service's Chief Arts Producer, now freelance.

REV: M. R. D. Foot in *The Spectator* (1 Oct 2008), 44.

LOC: Internet. Favorable.

REV: Hayden B. Peake in *Studies in Intelligence*, Vol.53, No.3 (Sep 2009), 41-42.

LOC: Internet. Favorable.

Shaffer, A[nthony] (1926-2001)
Sleuth. New York: Dodd, Mead, 1970, 125pp.

> A playscript. A miniature fictional masterpiece of the battle of wits where only one trickster survives. This was its first publication, preceding the British first edition by a year. The play premiered in 1970 in London at St. Martin's Theatre. It then opened on Broadway on 12 Nov 1970 where it ran for 23 months and won the 1971 best play Tony Award.

> Sometime around early 1971 I asked myself, "What would be the most elegant mystery story?" And, after brief thought, answered, "A story with only TWO suspects." I decided this would make an interesting basis for trying to plot a short story. Fortunately, before I got around to writing, I found I'd been preempted by a highly successful play currently on stages in both London and New York. It was Shaffer's Sleuth. Almost immediately, the surprisingly faithful movie version appeared in 1971, starring Lawrence Olivier and Michael Caine.

Shultz, Richard H., Jr. (1947-)
The Secret War Against Hanoi: Kennedy's and Johnson's Use of Spies, Saboteurs, and Covert Warriors in North Vietnam. New York: Harper Collins Publishers, 1999.

> A well-crafted account. Particularly relevant here are the sections on two clever and successful US military deception operations. The first, the Diversionary Program (codenamed FORAE), run by SOG's psywar section in late 1967, was designed to deceive North Vietnam as to the degree and nature to which it had been infiltrated by enemy agents (pp.58, 112-127).

Swinton, *Maj-Gen.* Sir Ernest
SEE: "Ole Luk-Oie" [penname]

Weiss, Richard J[erome] (1923-)
"A Touch of Monet," in Richard J. Weiss, *A Physicist Remembers* (New Jersey: World Scientific, 2007), 197-220.

An amusing fiction about deception & counterdeception in art forgery, one involving a games of wits between faker & fakebuster. See also "Ole Luk-Oie" (1909) and Whaley (1977).

Professor Weiss, after retiring from academia, had turned art fakebuster. This soon led to his writing this story. As he explained (p.196), writing in the third person:

> "When he met one of the scientists from Scotland Yard at an optical society meeting Richard's fondness for Sherlock Holmes sparked an idea. He created a character Robert Flowers, a Professor of Forensic Physics at King's College who consulted for Scotland Yard. He produced 12 short stories a la Conan Doyle and called the collection *A Dozen Forensic Flowers*. He sent them off to an agent in New York who enjoyed them, made a few suggestions, and tried to peddle them. No one bit."

So, on the Waste Not, Want Not Principle, he appended the first of these short stories to his autobiography.

Dr. "Dick" Weiss (1951 NYU PhD in Physics), an American, retired from King's College (London) in 1990. He developed an interest in art authentication in 1991, as outlined in his autobiography, *A Physicist Remembers* (2007), 243-248.

Whaley, Barton (1928-)
"The Door to Death: A Short Story of Counter-Deception." 1st draft, May 1977.

Printed in Bart Whaley, *A Reader in Deception and Counterdeception* (Monterey, CA: 2003), 397-403.

> When I wrote this piece, I'd assumed it was the first fictional story of military counterdeception. Only later did I discover I'd been preempted by 68 years by "Ole Luk-Oie" [pseudonym of Maj.-Gen. Sir Ernest Dunlop Swinton] (1909). While his story has a Know-Your-Enemy theme, my theme is one of provoking the enemy to verify your hypothesis. See also R. Weiss (2007).

Textbook of Political-Military Counterdeception: Basic Principles and Methods. [Washington, DC.]: Foreign Denial & Deception Committee, National Intelligence Council, Office of the Director of National Intelligence, Aug 2007, vii+186pp+CD.

LOC: BW.

A comprehensive overview.

Winks, Robin W[illiam] (1930-2003)
Cloak & Gown: Scholars in the Secret War, 1939-1961. Second Edition, New Haven: Yale University Press, 1996.

LOC: BW.

This reprints Prof. Winks' 1987 classic study, adding only a welcome new Preface and lightly correcting the original text — so lightly that it keeps the original pagination. Early but still enormously useful for its biographical and historical background details on Yale's contribution to American intelligence personnel, organizing, and practice during WW2. Particularly relevant are his strong chapters on Norman Holmes Pearson and James Jesus Angleton because both men had been deeply involved with WW2 double-agent work and Angleton also afterwards. Superbly researched throughout and intelligently analyzed. An extraordinarily rich source of deep insights on the craft of intelligence and counterespionage that reinforces my belief that all teams of deception planners and counterdeception analysts should include at least one historian.

Dr. Robin William Evert Winks (1957 Johns Hopkins PhD in International Relations) was a history professor at Yale since 1957, except 1969-71 when he was Cultural Attaché at the US Embassy in London. As a long-time devotee of detective fiction and espionage fiction & fact, he was almost unique in the high quality of his criticism of both branches of literature.

Young, Martin (1947-), and Robbie Stamp (1960-)
Trojan Horses: Deception Operations in the Second World War. London: The Bodley Head, 1989.

LOC: BW.

■ END ■

Volume II

WHEN DECEPTION FAILS:
The Theory of Outs

BARTON WHALEY

Editor: Susan Stratton Aykroyd

Foreign Denial & Deception Committee
National Intelligence Council
Office of the Director of National Intelligence
Washington, DC

August 2010

The views herein are the author's and not necessarily those of the Foreign Denial & Deception Committee

Volume II

WHEN DECEPTION FAILS:
The Theory of Outs

BARTON WHALEY

> When you put over a deception, always leave yourself an escape route.
> — Brigadier Dudley Clarke, as quoted in Mure (1980), 42

> If the performer knows enough outs, he will never fail to bring a trick to successful conclusion.
> — Jerry Mentzer, *Close-up Card Cavalcade*, Vol.2 (1974), 23

> In my experience, the 'out' has sometimes proved better than the intended effect.
> — Louis Histed, *The Magic of Louis Histed* (1947), 25

With pleasant memories of

Michael Handel

(1942-2001)

colleague and friend in
Cambridge (Mass.) & Jerusalem.

Contents

Introduction . . . . II-ix

Summary of Findings & Conclusions . . . . II-xi
    Main Finding. . . . . II-xi
    The 5 Causes of Failure. . . . . II-xiii
    The 5 Cures: The Rules of the Out . . . . II-xiii
    The 3 Ways to Minimize the Chance of Failure . . . . II-xv
    Conclusions: The 14 Lessons Learned . . . . II-xvi

Part One: Problems of Failure II-1

1. Causes of Failure II-1
    1.1. Design Failure II-3
    1.2. Initiation Failure II-4
    1.3. Transmission Failure II-8
    1.4. Inept Implementation II-8
        1.4.1. Target Doesn't Notice II-9
        1.4.2. Target Finds It Unbelievable II-9
        1.4.3. Target Judges It Irrelevant or inconsequential II-12
        1.4.4. Target Finds the Message Ambiguous II-14
        1.4.5. Target Misunderstands the Intended Meaning II-14
    1.5. Target Detects the Deception II-14
        1.5.1. Too Complicated: The Tangled Web Principle II-15
        1.5.2. Too Simple: The Too Perfect Principle II-18
        1.5.3. Target Succeeds at Deception Analysis II-22

2. Degrees & Types of Failure II-24
    2.1. Partial Failure II-27
    2.2. Total Failure II-31
    2.3. Backfire II-46
    2.4. Turnaround vs Turnabout: Blunder vs Entrapment II-56
    2.5. Aborts II-60
    2.6. Pseudo Failure II-65
    2.7. Potential Failure II-69

3. Coping in Military Practice II-73
    3.1. Soldier's Myths II-73
    3.2. Soldiers' Practice II-74

Part Two: Solutions to Failure II-77

4. Coping in Theory II-77
   4.1. Toward Theory II-77
   4.2. The "Blow-Off" Technique II-78
   4.3. The Theory of "Outs" II-79

5. Military Application of Outs II-81

6. Priorities for Outs II-83

7. Beyond the Out: Asymmetries between Deceiver & Target II-86

Bibliography II-91

Appendix: Extracts from 1982 CIA contract paper titled *Deception Failures, Non-Failures and Why* II-98

Indices II-105
   Index of Cases—In Order of Appearance II-105
   Index of Cases—Chronological II-108

Introduction

The title page frames the problem of deception failure. It opens with the main title that implies the question of what can we do "when deception fails." It ends by suggesting the best answer, namely the Theory of Outs. This answer is stated both in the subtitle and in the title page's three quotations. Their authors are three magicians—an American pro dealer & publisher of magic, an English customs duty collector & creative amateur conjuror, and an English amateur magician who happened to be Britain's most inventive and consistently successful military deception planner and operator in WW2. These highly experienced authorities give the answer to this bothersome problem of actual or potential failure of our "best laid schemes." This is The Theory of Outs. That's magicians' jargon for the various ways to recapture the initiative whenever a trick a deception—fails. That is the subject of Chapter 4.3.

The Theory of Outs is the most practical guideline for getting us out of all types of failed deception operations. But how do we get into these potentially disastrous situations in the first place? That is the subject of Chapter 1.

This study sought to collect, report, and critique all publicly known examples of military deception failure. Because few combat deception failures are documented and the topic is virtually unexplored, the research was extended to include a few similar failures in such closely related fields as camouflage, covert operations, psychological operations (including "black propaganda"), and counter-espionage. Moreover, it looks not only at the very few extreme or nearly complete failures but also at those that backfired, were aborted, or had mixed results. In addition it criticizes a couple of examples that, while successful, courted unnecessary failure. Taking this broad view of the subject yielded a total of 60 case studies.[1]

Note that this set of cases comprise an "opportunity sample" only. Consequently, they have substantial anecdotal value but, strictly speaking, are not appropriate for statistical analysis. The paper ends with an explanatory theory that pinpoints those steps the deceiver must take if success is to be rescued from impending failure.

■ ■ ■ ■ ■

---

1   I conceived and began preliminary research on this study sometime in the mid-1980s. By 21 Oct 1987 an early draft of 52 pages listed 47 case studies.

Finally, note that this study excludes the factor of self-deception as irrelevant to our attempt to understand how and why deceptions of another party can fail. Some studies have included this factor.[2] But that unnecessarily confuses the terms "self-deception" and "other-deception". A simpler model of perception that shows this relationship:

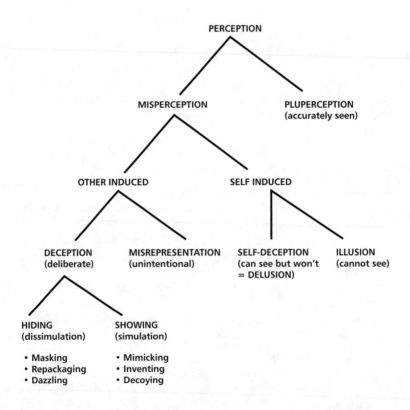

---

2   For example, Maxim (1982), 50-51. That study gave (pp.10-11) as a case study, the Jul-Oct 1940 "Battle of Britain" where Luftwaffe Intelligence grossly underestimated the capabilities of British radar and fighter aircraft. That study also cited (p.11) the USAF's similarly self-satisfying underestimation in 1976 of the capabilities of the Soviet MiG-25 (FOXBAT) fighter despite having dismantled & studied a complete model flown out of Russia by a defecting pilot. See *Wikipedia*, "Mikoyan-Gurevich MiG-25" (accessed 20 Mar 2010).

Summary of Findings & Conclusions

What happens when deception fails? Under what circumstances and with what consequences? And, most importantly, what can be done to avoid such failures in the first place and to minimize damage if it happens?

This paper answers these questions. It is the most comprehensive overview of why military deception operations fail. Moreover, it is the first paper to identify a general solution to strategic, operational, and tactical military deception failure—The Theory of Outs.

Surprisingly, given the obvious importance of deception failure, only a single previous study, Maxim (1982), even attempted to treat it as a general problem.[3] That unclassified CIA-sponsored study, now nearly three decades old and based on only 11 case studies, received almost no attention. It quickly vanished into an archive, its very existence limited to two obscure bibliographic references. One of these led to the CIA Library's copy, which I then accessed in the late stages of research on this paper.[4]

Consequently this new study not only revives this important and surprisingly overlooked problem, but explores it to new depths. And, because this new monograph is anchored by 60 case studies (including the 11 earlier ones), it has a much stronger data base than the previous study, one that has permitted a fairly deep analysis.

That deeper analysis has enabled this study to identify and present a useful general solution to failure that has long proven highly effective in two quite different professional domains. The first is the confidence artist's Blow-off or Cooling Out the Mark. But that is, at most, a rough principle. A far more developed theory and systematic practice is the professional magician's Theory of Outs. It will be this specific conjuror's model that will be applied here.

Main Finding

All deception plans are subject to failure. That is obvious. However, such failures are rare in practice. Nevertheless, their ever-present possibility is a concern. Some commanders and military planners become so concerned that they tend to discount, prematurely abandon, or even entirely avoid deception

---

3   The short account of deception failures in *FM 90-2 (Battlefield Deception")* (1988) was simply copied (without credit) from Maxim (1982) along with Maxim's earlier paper titled "Deception Maxims" (1981).
4   My source was WorldCat.org. The other citation was A. Zaklad & others, *Deceiving the Opposing Force* (1990), a paper prepared for the US Army by staff at Ft. Huachuca.

plans. This is not cost effective.[5] Fortunately, certain identifiable steps can and therefore should always be taken to anticipate deception failure and minimize its consequences.

Perhaps the single most important finding of this study is that the deception planner cannot work alone to improvise the steps needed to overcome a failed deception. The Commander must be involved and her involvement should begin with the planning of the initial deception. Specifically, the Commander must, at the minimum, build a plausible alternative goal into the initial overall operational plan. The deception planner needs this input as the basis for designing one or more "Outs".

This finding has at least one major implication. It gives the most potent argument for assigning deception planning to the Plans staff rather than to Intelligence, as has often been the case. But having the deception planners in Plans greatly facilitates the interaction between commanders and their deception planners to assure the presence of "alternative goals" in every operational plan. In the best case, deception planners in Plans are thus better situated to learn their commander's alternative goal or goals and the priorities she assigns them. If, as often happens, commanders neglect to articulate their alternative goal or goals, deception planners in Plans are better placed to get them or their staff to make the alternatives explicit. And in the worst case, if the Commander has neglected to even think of any contingent alternative, the deception planners are in a better position to urge that this be done and to contribute positively to that end.[6]

An important consideration is that deception rarely yields either/or results—total success much less total failure. The result is almost always somewhere between. Moreover, failure in one mode (such as the place of attack) may be balanced by success in another such as the time, strength, or style of attack). Also, should the *strategic* deception fail, the *tactical* one may succeed. Finally, if several places are threatened, at least one feint may succeed. Analysts of deception operations should learn to avoid simplistic labels of either "success" or "failure" and adopt as sensitive a scale as their data permit.[7]

Finally, we can bury the old myth that the risks and costs of deception failure are so great that it's sometimes better to have no deception at all. In fact, deception failure is rare and total failure is very rare. Moreover, failed deceptions cause no

---

5   As shown in Whaley, "The one percent solution: costs and benefits of military deception," in John Arquilla & Douglas Borer (*editors*), *Information Strategy and Warfare* (New York: Routledge, 2007).
6   This point is also stressed by Major (USA) John Arbeeny, based on his experience in 1983-85 running deception planning in Korea where deception was handled as a staff function under G-3 (Ops).
7   For example, I found in 1969 that a 5-point scale worked quite well with even the relatively crude data for the 168 cases in my database at that time.

more harm than no deception; and any deception—however crude—is better than none. In sum, while the best laid plans of deceivers often go askew, with proper advance design, the deceiver can always have at least one "Out".

The 5 Causes of Failure

Deception fails only in the following five general circumstances:

　　1) Design Failure

　　2) Initiation Failure

　　3) Transmission Failure

　　4) Inept Implementation

　　5) Target Detects the Deception

In most historical cases, the deception game *ends* at whichever of these five main steps was reached first. However, a persistent deceiver can, *if prepared*, always continue the game one stage further.

This study collected, researched, and analyzed 60 cases of various types and degrees of "failure". The main conclusions are presented as 14 "Lessons Learned".

The 5 Cures: The Rules of the Out

Always have an "Out". This requirement has five specific components:

Rule 1: Always preplan your Out. This is the responsibility of the deception planner or planners, who must be ready to act instantly when deception fails. It is not the responsibility of any other unit, coopted to "put out the fire".

Rule 2: Tailor each Out to fit each deception plan. There is no Users Manual of Outs, no handbook, no checklist. Such detailed and specific sets of contingency plans, sitting on a nearby shelf ready to be consulted in an emergency, are effective and appropriate for commercial aviation pilots.[8] Conversely, back-up plans must be both kept up to date and tailored to specific local conditions. We were reminded of this recently (June 2010) on learning that BP's 583 page on-the-shelf oil spill manual for its Gulf of Mexico off-shore wells had been deficient on both counts.[9] But neither such type of *general* emergency handbook will be effective or even applicable to a deception plan gone wrong.

8　Atul Gawande, *The Checklist Manifesto: How to Get Things Right* (New York: Holt), 2009, 114-135.
9　BP, *Regional Oil Spill Response Plan: Gulf of Mexico* (Revised: 30 Jun 2009).

Rule 3: Although the Out will be designed after the deception plan, it must be in place and primed to activate before the actual deception operation is launched.

Rule 4: At most, in those larger scale operations where fairly large staffs and longer-range planning times are available, gaming or even simple "brainstorming" or "devil's advocate" sessions can be effective ways to identify likely snags & outs. At the minimum, planning for the Out can be the task of a junior member of the planning team. But, whatever the scale of operation, an Out must be part of every deception planning checklist.

Rule 5: Finally and obviously most importantly, always seek to minimize catastrophic consequences—worst case scenarios. We do this by building at least part of the Out into the plan and operation. The clearest precedent in the military deception context was set in WW2 by the British WW2 Double-Cross Committee.[10]

- **The problem:** How can we preserve the credibility of any given deception channel after the enemy realizes that channel had been giving them false information? Take the example double agents Some doubles can be sacrificed and often are. But that is a costly option when you have few disinformation agents or few other channels.

- **The solutions:** To preserve any deception asset (channel) always give them at least one of four main types of potential Outs—ones that are made known to the enemy from the first:

    Solution 1: The simplest Out in espionage relies on the old principle of Never Send a Novice to do an Expert's Job. Ideally, your enemy has employed only experts who report information appropriate to their expertise—the rocket engineer reports on rocket development, the HQ personnel clerk reports on her unit's strength, equipment, readiness, location, etc. Such information from such sources tends to be highly reliable. However, if that expert, for example the HQ clerk, reports that the rocket's range is such-and-such miles, there is much room for simple misunderstanding and misinformation.

    Solution 2: Another simple Out is to make your individual messages sufficiently vague—open to more than one interpretation. Then, that, after the enemy analyst realizes he had picked a wrong implication, he will blame it on the inherent ambiguity of the information and not on you. Remember, the good deceiver never spells out the implications of the data. Instead, the good deceiver

---

10   The most detailed and comprehensive account is Terry Crowdy, *Deceiving Hitler: Double Cross and Deception in World War II* (Oxford: Osprey, 2008).

leaves the analysis and interpretation of the data to the enemy's intelligence analysts. Consequently, they are more apt to blame themselves than blame you for having drawn a false conclusion from ambiguous data.

Solution 3: If your agent (or other type of source) must be blown, make it one of your real agent's notional sub-agents. Then, if the sub-agent is revealed as a misinformed incompetent, a greedy pretender to having secret information for sale, or even a turned agent, only marginally damage accrues to your main agent's credibility. Indeed, credibility will actually be enhanced, if the agent has herself previously raised doubts about the competence, greed, or loyalty of her sub-agent. This puts the onus for having accepted the sub-agent's information back on the deception victim's intelligence service. This is a surprisingly repeatable technique.

Solution 4: The Delayed Message Trick is risky but effective. It is also rare and must remain so. You transmit a complete and genuine warning of your impending operations plan but do it through a channel that you know will be too slow for the enemy target to effectively react. Do this well (as in the GARBO D-Day case)[11] and the enemy will blame the known inadequacies in his own communication system.

The 3 Ways to Minimize the Chance of Failure

There are three ways to reduce the probability of a deception operation failing. One way to minimize the likelihood of failure—the usual and usually inadequate way—is to anticipate the developing situation. Briefly put, this involves nothing more than considering and evaluating the various alternative hypotheses. We already have ample tools for doing so—more-or-less powerful techniques such as gaming, Devil's Advocacy, brainstorming, and the recent Heuer-Stech Analysis of Competing Hypotheses (ACH). I won't go further into that topic but go directly into the general principles—the do's & don'ts—that apply to both these ways to minimize deception failure.

The second, and usually more effective way to avoid failure is by understanding and acting on the lessons learned from the failure of others. That is the main product of this paper.

A third way to avoid failure is the subtly psychological game that play upon your opponent's ignorance of just how efficient or inept you are at deception. This, of course, is part—an important but largely overlooked part—of the old

---

11   Whaley, *Turnabout* (FDDC: Jan 2010), Case 6.1 "GARBO".

"Know Your Enemy" admonition of Sun Tzu and Polybius. However it involves a role-playing game that has only been systematically developed, perfected, and used by magicians. Indeed, only consciously and systematically so by the more sophisticated practitioners, those most aware of the vulnerabilities of each specific audience they confront. This involves that the deceiver adopt the role that best fits the specific target audience. This important method for significantly reducing deception failure is explored—for the first time—in Chapter 7 ("Beyond the Out").

This paper's array of cases where deception failed, aborted, or ran an unnecessarily high risk of failure. That is a sobering record which casts a harsh light on the abilities of the individual deception planners involved in these fiascos and potential fiascos. It does not, however, throw deception planning theory itself into disrepute, because in each case it was the planners who, through inexperience or incompetence, had failed to understand and practice the guiding principles of their art.

Ironically, it is fortunate that so many of our colleagues have failed or proven less than perfect in this art of deception. True, it is an embarrassment to our professional pride. True also that examples of past failure make it more difficult to sell the concept of deception to our Commanders. But a saving grace is that these case studies of partial failure can be a more fruitful source of "lessons learned" than the many case studies of success. And, if we are prepared to learn from the past blunders of others, we may better avoid future ones ourselves.

Success tends to go unquestioned. "Don't fix it if it ain't broke" is a wise rule of thumb, but one that should never be applied mindlessly. Just because a particular deception operation like "The Man Who Never Was" succeeded is no reason to avoid close criticism to uncover potential flaws.

Conclusions: The 14 Lessons Learned

Failure generates a rich array of lessons learned. Those that emerged from this survey are summarized now. If some of these "don'ts" strike you as so obvious as to seem like mere common sense, remember that each was a hard-earned lesson from at least one professional deception planner's failure. I find it odd that the cliche about being "wise after the fact" is always used in a contemptuous sense. But, as Brigadier Bernard Fergusson (Lord Ballantrae) wrote of his first and only failed battle:[12]

> Not to be wise after the event is stupid: not to try to be, criminal. Only by such wisdom does one learn; and after a campaign one

---

12   Bernard Fergusson, *The Wild Green Earth* (London: Collins, 1952), 99. Re Case 2.3.3.

should rehearse and recognize one's mistakes, for the profit of others as well as one's self.

On the sound teaching principle that "do's" are more memorable than "don'ts", the following 14 lessons highlighted in this study are all stated here in positive form:

1. First and foremost, as a deception planner, know your Commander's intentions and battle plans. These set the basic constraints within which your part of the planning will take place. The deception planner should view all such constraints and limitations as a challenge to be surmounted and not as cause for despair. (Case 2.1.1)

2. Second, try to find out what your Commander wants the target to do, that is, how he wants the enemy to *react* and not merely what he wants the enemy to *believe*. If, as happens, the Commander doesn't grasp this concept, the deception planner must infer the answer from the Commander's operations plan. (Case 2.1.1)

3. Know the tools of your craft, that is, what deception assets—human as well as materiél—are available and their capabilities and limitations. Again, the inevitable limitations are a challenge to overcome this "option of difficulties". Moreover, experience can teach the need for a new asset that can perhaps be developed in time for later operations. (Cases 1.2.1 and 2.3.4)

4. Conversely, let your clients—your commander, parallel staffs, and subordinate commands—know *your* capabilities, what you can and can't do for them. Remember that deception personnel can only *simulate* the false; it is up to the real units to *dissimulate* themselves, and they can only do this if properly indoctrinated and trained in such OpSec dissimulative techniques as radio silence and camouflage discipline. (Case 1.2.1)

5. Coordinate the plan with any other deception planning groups or operational units whose cooperation is needed. The higher echelon planning units must set limits on the lower, such as which cover targets to avoid, to prevent working at cross purposes. (Cases 1.2.1; 2.3.2; and 2.3.3)

6. Similarly, notify in advance the HQs of all friendly units that could be adversely affected by the operation if they fell for the deception. Feigned withdrawals or apparently alluring gaps at the junction with a neighboring unit might provoke that unit's commander to begin his own withdrawal or to extend his line to fill the "gap". (Case 2.3.3)

7. Tailor each plan to take advantage of the target's military doctrine, capabilities, intelligence SOP, and preconceptions. As Sun Tzu advise:

"Know your enemy and know yourself and in a hundred battles gain a hundred victories."

- a. Doctrine. Use the rigidities in your target's military doctrine against him. This is the Play By Your Own Rules principle of asymmetric warfare.
- b. Capabilities. Play upon the target's strengths and weaknesses. Find his vulnerabilities.
- c. SOP. Knowledge of the target's intelligence standard procedures tells you what channels are open to him and what types of ruses have better chances of being sold. One of the many ironies in this game is that deception works best against an alert, perceptive, and efficient enemy intelligence service. The better he is at his job, the easier it is to con him. Sophisticated deceptions will be wasted on incompetent intelligencers. (Cases 1.2.1; 2.2.3; 2.2.4; 2.2.12; 2.3.5)
- d. Preconceptions. Wherever possible, adopt a plan that plays upon and reinforces the target's preconceptions. While it is possible to sell deceptions that run counter to the target's preconceptions, it takes longer and is less often successful. (Cases 2.1.2; 2.2.5; 2.2.14)

8. Keep it simple. Avoid plans that by their nature require complicated elements; and resist adding unnecessary complications to an already elegant plan. Complex ruses such as "The Man Who Never Was" require a seamless fit of all components and closest coordination with operational units, an ideal situation that can easily break down. If the target detects a single discrepancy, the whole operation is likely to unravel; and complex plans, by adding more discrepancies, increase their chances of being detected. Unnecessarily multiplying the opportunities for failure creates the ideal environment for a "perfect storm" of unmasking clues. Experienced magicians avoid "overproving", having learned the hard way that it only draws unwanted suspicion. (Case 1.5.1.1)

9. Make plans within plans. This does not contradict the "keep it simple" rule, because it refers to separate deception sub-plans for lower level units, each of which can be kept simple. (Cases 2.1.1; 2.1.2)

10. Fit the plan to the time available, whether the D-day or H-hour deadline. An abort will occur unless the plan can become operational in time to serve its purpose. (Case 2.5.5) Also realize that a few impatient Commanders tend to unexpectedly shorten their deadlines. Thus, in North Africa in WW2, by snap decisions to attack prematurely did an impatient German Field-Marshal Rommel twice surprise Berlin and the British Eighth Army—

which was reading the Rommel-Berlin radio traffic. In such cases, it can be later than you think.

11. Keep it short. That is, keep the actual deception operation as brief as possible. The longer it lasts, the greater opportunity for the target to detect it. And realize too that most commanders tend to extend their deadlines. Then, it may be earlier than the victim thinks.

12. Seek feedback. Accurate and timely feedback gives the only clues that the operation is succeeding or failing. Strategic deception plans may never again enjoy the quality and coverage in feedback that the British got from ULTRA in WW II. Tactical deception plans, however, do not need such sophisticated technical means. Conventional combat intelligence methods (direct visual and electronic observation, POW interrogation, etc.) is often enough to verify that the target has "bought" the deception. (Cases 2.2.3 and 2.2.10)

13. Depending on the feedback content, fine-tune your operation. (Cases 2.2.2 and 2.2.10)

14. Last but not least, stay flexible. Design your plan to include at least one "out". This requires prior knowledge of your Commander's intentions and operational plans and brings us around full-circle to Point 1. (Case 2.2.1.)

PART ONE:
Problems of Failure

This first main part elaborates on the findings in the previous Summary. Additionally, it draws in the documented evidence, which is a combination of the case studies collected for this monograph and other sources.

1. Causes of Failure

> The best-laid schemes o' mice an' men
> Gang aft agley
> An' lea' us nought but grief an' pain,
> For promis'd joy!
>
> — Robert Burns, "To a Mouse" (1785), Stanza 7

Deception fails only in the following five general circumstances. Taking into account some useful sub-categories these five are:[13]

1. Design Failure

    No design—pure improvisation

    Design omits simulation—limited to "security" (dissimulation)[14]

    Design omits outs

2. Initiation Failure

3. Transmission Failure

4. Inept Implementation

---

[13] Barton Whaley, "Toward a General Theory of Deception," *The Journal of Strategic Studies*, Vol.5, No. 1 (March 1982), 189, gave only the four categories of "Target takes no notice of the intended effect; Target notices but judges it irrelevant; Target misunderstands its intended meaning; and Target detects its method." The fifth category, "Deception material fails to reach the target," was added in the 1987 draft of the present paper.
Douglas Webster, then a Lieutenant (USN). preferred a two-tier model: 1) Ineffective Communication (where my 2nd, 3rd, & 4th circumstances were sub-sets) and 2) Detection. BW conversation with Webster, 19 October 1987.

[14] True deception requires simultaneously dissimulating (hiding the real) and simulating (showing the false). "Security" measures—which typically focus narrowly on secrecy (hiding)—invite unwelcome attention.

> Target doesn't notice
>
> Target finds it unbelievable
>
> Target judges it irrelevant or inconsequential
>
> Target finds the message ambiguous
>
> Target misunderstands the intended meaning

5. Target Detects the Deception

> Too complicated. The Tangled Web Principle
>
> Too simple: The Too Perfect Principle
>
> Target succeeds at deception analysis

In most historical cases, the deception game *ends* at whichever of these five main steps was reached first. However, a persistent deceiver can, *if prepared*, always continue the game to the next stage.[15] To elaborate:

Effective deception design or planning must anticipate all of these contingencies. A wise deceiver will seek feedback by monitoring the target's responses to confirm that these contingencies are being met. If feedback shows the existence of either circumstance 2, 3, or 4, then the deception planner should make appropriate adjustments. For example, the planner can seek one or more channels that are open to the target, try to turn up the "signal" level, make the indicators more relevant, or fine-tune them to reduce misunderstanding.

However, if the target actually detects the deception's M.O., the planner has lost the game—*unless* she has prepared a completely different, alternative method, ready to be plugged in at that crucial point. Remedies available to the frustrated deceiver come under the head of what magicians call the Theory of "Outs".

If the deceiver's deception is unmasked *and the target doesn't know this*, it then becomes possible for the detector (original target) to plan and carry out his own deception operation to confound the original deceiver. Their roles reverse and the target of deception becomes the deceiver and the quarry becomes the hunter. This rare but important type of deception that pits one deception against another has gone by several terms including even "counterdeception", but it deserves its own terminology. In a recent paper, I recommended the terms turnabout (as in "Turnabout is fair play.") or, as a synonym, double-cross. I propose we use the terms turnabout (as in "turnabout is fair play") or double-cross.[16]

---

15 As analyzed in Whaley, *The Maverick Detective; or, The Whole Art of Detection* (manuscript in progress since 1986).

16 See Whaley, *Turnabout: Crafting the Double-Cross* (FDDC, Jan 2010).

Having detected the deception, the detective can:

- Pretend to not notice the deception's intended effect or method. If the deceiver has accurate feedback on this, he will be forced to return to the drawing board to design a new deception or waste more effort in reinforcing the old one.
- Pretend it is irrelevant, with the same result as above.
- Pretend to misunderstand the effect. Here, the counterdeception game begins in earnest, if the detective presents what is "misunderstood" as something the enemy doesn't want him to believe. This will not only force the enemy to drop the original deception but may even pressure him to abandon or modify his original real plan.
- Pretend to have detected more of the deception than is, in fact, the case or pretend to have not detected it at all. If the detective chooses the first option and exaggerates his achievement, the deceiver will react as in the above condition. However, if the detective conceals his detection, then he can set his own trap and bite the biter.

Let's begin our survey of deception failures by looking at two marginal categories—the failure to even attempt it and failures in transmitting the deception material.

1.1. Design Failure

The Problem:

A weak, ill-conceived and ineptly designed deception plan greatly reduces the likelihood of the operation's success. An obvious point, but too often improperly handled.

The Solutions:

- Deception planning staffs should be embedded in the same organizational structure as regular planning staffs. This assures that each deception plan will be:

    a) Vetted for possible collateral damage by other plausibly involved parties (such as psyops, CI, CE, diplomats, allies, or air, naval & ground units); and

    b) Approved by higher command, which may have its own reasons for vetoing. Even when time is of the essence, effective vetting can be expedited, particularly when the deception planners are sufficiently sophisticated & farsighted to include staff members of these other units in their later planning stages.

- War gaming of deception plans will often be the most powerful tool for spotting the weak, counterproductive, or even fatal parts. Again, given a small permanent gaming infrastructure, such role-playing simulations can be run in a couple of days, as the Israeli PM's office proved in preparation for the 1976 Entibbe raid.

The three most common types of design failure are:

- No design for deception—pure improvisation. Historically, this has been the usual case. It should never be the case.
- Design omits simulation—limited to "security" (dissimulation). Properly executed deception requires simultaneous use of dissimulation (hiding the real) and simulation (showing the false). "Security" measures that focus narrowly on secrecy (hiding) invite unwelcome attention. Again, this is the usual case but need not be.
- Design omits outs. And yet again the usual but unnecessary situation.

1.2. Initiation Failure

The Problem:
Deception can be part of the soldier's kit—perhaps somewhere in that pile of field manuals—but he simply forgets or doesn't bother to use it. Tight camouflage discipline and effective deception measures are common enough in the American forces among experienced combat soldiers (i.e., survivors). This is true for individuals to small tactical units such as 2-person sniper teams and Special Forces teams. However, it becomes quite rare as we move up the scale into operational and strategic operations.

The Solution:
Emulate the "best practice" of experienced deceivers. Do what they do, although this will usually require improvisation. This will, with rare exceptions, require disregarding one's field manuals and all other non-deceptive SOP.

What are the consequences of having an approved or SOP deception plan in place but not implementing it? In other words, not vetoing or aborting such a plan by design or change of circumstances that render an old plan obsolete (as discussed in Section 2.5) but by simply neglecting to use it? Although such sheer forgetfulness, while rare, does occur, it is apt to yield unacceptable loss, as seen in the following cautionary tale.

CASE 1.2.1:
RAF Bomber Command Forgets to Use Deception, Germany 1944

On the night of 30/31 March 1944 RAF Bomber Command launched a major raid on the German city of Nuremberg. This raid was resulted in an unusually high rate of losses to Bomber Command. Of the 795 bombers involved, 84 were lost. This loss rate of nearly 12% was more than double the usual and considerably more than the replacement rate for aircraft and crews. At this higher loss rate, Bomber Command would soon run dry. This raid was unique only in that the bomber stream was sent into enemy airspace without the protection of any deception measures.

None of the several existing, conventional, and effective deception plans were carried out. In this case the cause was an administrative blunder—apparently no one notified the deception units. The immediate consequence was an unacceptable—that is, unsustainable—loss rate to the total British bomber force. But, because this was what social scientists call a "natural experiment", it gave nearly conclusive proof of the overall effectiveness of the deceptive measures in keeping bomber loss rates within an acceptable range.

This one raid was typical of all other Main Force operations during the so-called Battle of Berlin, which comprised 35 raids on cities throughout Germany during the four-and-a-half months from mid-November 1943 through the following March. That one raid was typical in all major particulars except one—deception. Strangely, Bomber Command omitted virtually all its best ruses. The Main Force of 795 heavy bombers flew an almost direct course toward their target, with only two minor turning-points. None of the normal diversionary operations were undertaken. (Later, Bomber Command limply attributed this to "conditions over the North Sea" without explaining why other routine operations were able to proceed there.) In any case, a mine-laying raid by 50 Halifax bombers in the Heligoland Bight did not confuse the German ground controllers. Nor did the small-scale spoof raids on Kassel, Cologne, Frankfurt, and other minor targets, because the Germans immediately recognized the raiders as unthreatening Mosquito light bombers.

A German interceptor force of 246 night-fighters was scrambled and waiting when the Main Force crossed into German airspace at midnight. In the clear night-sky, lit by a half-moon, the interceptors could see the bomber stream, so they no longer needed their ground controllers' instructions, which in any case were now being jammed by the bombers' on-board jammers. During the hour-and-a-half the bomber stream was fighting its way the 250 miles into Germany, the night-fighters shot down about 58 bombers. Given the furious distraction of the aerial combat and the high tail winds, about 27 other bombers aborted or lost their target and the rest were so scattered that the ground controllers did not actually identify Nuremberg itself as the actual target until two minutes before the attack began by the 710 remaining bombers. This element of fortuitous surprise did catch seven squadrons of the single-engined "Wild Boar" interceptors hovering outside Berlin and Leipzig. However 21 squadrons (75% of all scrambled interceptors) were clinging to the bomber stream, destroying another 22 over Nuremberg and a final 15 on the homeward flight. In all, the British raiders suffered 94 bombers lost, an appalling 11.8% rate. In addition, 12 bombers were damaged beyond repair and 59 moderately damaged. Nuremberg received only minor bomb damage.[17]

If we simply take the British Official History statistics and recast them as in the following table, the key point is made.

British Heavy Bomber Losses in the Battle of Berlin, 18 Nov 1943 – 31 Mar 1944

Target	No. Raids	No. Sorties	No. Lost	% Lost
Nuremberg	1	795	84	11.8
Berlin	16	9,111	492	5.4
All other targets in Germany	18	10,318	461	4.5
TOTALS:	35	20,224	1,047	5.2

NOTE: During this same period, Mosquito light bombers flew 2,034 diversionary sorties, losing only 10 aircraft (0.4%).[18]

Three points are clear:

- The confusion and surprise given by effectively applied deceptions yield average loss rates of only about 5% (and those other 34 raids included some where, despite good deception, the German ground controllers or some interceptors accidentally guessed the real target).
- Where deception was poor and surprise was slight, the loss rate soared to nearly 12%.
- Costs of the electronic and spoofing operations themselves were small relative to their effect on the Main Force efficiency and loss rates.

---

17 Martin Middlebrook, *The Nuremberg Raid* (New York: Morrow,1974); Sir Charles Webster & Noble Frankland, *The Strategic Air Offensive Against Germany, 1939-1945*, Vol.2 (London: Her Majesty's Stationery Office, 1961), 192-193, 207-209; Alfred Price, *Instruments of Darkness* (London: Kimber, 1967), 196-198; and *The Royal Air Force*, Vol. 3 (London: Her Majesty's Stationery Office,1954), 28. See also Cajus Bekker, *The Luftwaffe War Diaries* (Garden City, NY: Doubleday, 1968), 339-340; David Irving, *The Destruction of Dresden* (London: Kimber, 1963), 54-55; and Anthony Verrier, *The Bomber Offensive* (London: Bratsford, 1968), 293.

18 Webster & Frankland, Vol.3 (1961), 199.

This period (November-March) was one in which the normal 5.2% attrition rate in heavy bombers was "acceptable" to Bomber Command, because replacement of planes and crews could keep up—indeed the total daily average number of available heavies (with trained crews) increased from 864 to 974 over those five months. However, any attrition rate averaging over 5.7% in that same period (i.e., costing an additional 110 planes) would have been unacceptable. In other words, the already narrow margin of operational existence of Bomber Command in that period should almost certainly be entirely credited to the deception planning, which was then "saving" about 7% (give or take ½%) of the sorties, or at least 1,300 heavy bombers—more than the total original force.[19]

It is commonly argued that the strong needn't bother with deception. True, but only in two circumstances—both illustrated by this case. First, in the short run, if the stronger party isn't bothered by its own body bags. Second, in the long run, if it isn't concerned that the attrition rates from successive battles will exceed the rate of replacements.

CASE 1.2.2:
British Black Propaganda in the Suez War 1956
A sufficiency of blunders.

As part of the disastrous Anglo-French attempt in Oct-Nov 1956 to seize the Suez Canal and bring down the Egyptian government of Abdul Nasser, R.A.F. bombers tried to knock out Radio Cairo. I'll leave this story to the retired WW2 remarkable chief of "black" propaganda, Denis Sefton Delmer:[20]

> I am told that the plan behind this operation was for a British-controlled Arab radio operated from Cyprus to take over the Cairo frequency and broadcast anti-Nasser programmes on Nasser's wavelength. The plan came to nought however for two excellent reasons: (1) The R.A.F. knocked out the wrong radio, and (2) the Cyprus Arabs—no one apparently had thought to make sure of them in advance—came out on strike led by their British director when the British military took over the station.

Here was a rare case where everything was bungled. The precision bombing surprise strike on a wrong uncamouflaged & undefended target was inexcusable but not unique—in 1999 in Belgrade the CIA would manage a direct hit with 5

---

19 Price (1967), 243, estimates "more than one thousand [RAF] bombers and their crews" saved from December 1942 to October 1944 by ECM alone. During that period 4,490 RAF bombers were lost on night raids. James Phinney Baxter, *Scientists Against Time* (Boston: Little, Brown, 1948), 169, reports an official American estimate that "radar counter-measures saved the United States Strategic Air Force based on England alone 450 planes and 4,500 casualties."

20 Sefton Delmer, *Black Propaganda* (London: Secker & Warburg, 1962), 296.

smart bombs on a misidentified target (the Chinese Embassy). But even worse was the failure of the British psywarriors to properly vet their own coopted civilians in Cyprus. Obviously, the British deceivers of 1956 had lost that keen edge exhibited just 11 years earlier by their predecessors—Jones, Clarke, Bevan, Masterman, & Co.

1.3. Transmission Failure

> I shot an arrow into the air, It fell to earth, I knew not where; For, so swiftly it flew, the sight Could not follow it in its flight.
> — Longfellow, "The Arrow & the Song" (1845)

The Problem:
Deception will fail in transmission between deceiver and target in two obvious conditions—mechanical or human.

First, if the channel through which the deceiver attempts to communicate its spurious signals ("sprignals") does not exist, does not connect to the target, or is temporarily disconnected. A possible example is Case 2.2.2, a certain one is Case 2.2.4.

Second, if one or more of the human gatekeepers along that channel delete, garble, or misunderstand the sprignals. A likely example is Case 2.2.3.

The Solution:
To prevent this problem the deceiver must have at least some general knowledge of the technical and human components of the opponent's intelligence system—some understanding of that system's strengths & weaknesses.

1.4. Inept Implementation

The Problem:
Too many otherwise clever deceptions risk failure through sheer incompetence in planning and/or execution.

The Solution:

The only remedies are sound theory-based teaching and much realistic training or field experience.

The four main results of inept implementation are:

1.4.1. Target Doesn't Notice

The Problem:

The target simply takes no notice of the dangled effect.

The Solution:

Either turn up the volume or rephrase the message to make it seem salient in terms of the target's value system.

One set of examples is seen in the nearly inevitable failure in WW2 of British CE "coat-trailing" of double-agent candidates before German Military Intelligence (Abwehr) recruiters, as seen in Case 2.2.11. In that case, it took a sequence of these seemingly tempting dangles for the British to realize that their efforts were ineffective—evidently the Abwehr agent recruiters preferred to develop their targets from scratch.

1.4.2. Target Finds It Unbelievable

The Problem:

The Target notices & understands the message but simply judges it unconvincing.

The Solution:

Know your enemy, your target audience. Learn to distinguish what things they tend to accept on faith, those that will fit into and play upon their preconceptions, and those they can be led to accept.

CASE 1.4.2.1:
Operation: Hitler's Big Lie That Flopped, 1939

A diabolically cunning provocation rendered unconvincing by clumsy handling.

> Heydrich ... [was] saying, "Ah, yes! That's why war began." No one was fooled.
>
> — André Brissaud, *The Nazi Secret Service* (1972), 237

Hitler planned his unprovoked invasion of Poland for late August 1939. But he wanted it to seem to all concerned that it was the Polish government who had provoked & justified his action. This deception was assigned by Himmler's SS to the Nazi Party's intelligence & security organ, the Sicherheitsdienst (SD), headed by Lt.-Gen. Reinhard Heydrich. The operation would be essentially pure psychological warfare.

Two sets of incidents were planned. Both were of the "false flag" type in that they had German intelligence & security officers staging warlike incidents on Germany soil—specifically in Upper Silesia near the Polish border—while making it seem they were coordinated attacks by small Polish Army units. The first type of incident comprised simulated raids on two Silesian border sites—a customs house at Hochlinden and a forestry office at Pitschen. The second type of incident was much more elaborate and openly inflammatory. It simulated both the seizure of the Silesian radio station at Gleiwitz and the nationwide broadcast from there of a message denouncing Hitler as a warmonger.

The prime rule of all plausible deceptions require that they have at least some corroborative evidence in depth, ideally over a second channel ("multiple source intelligence"). In this case, almost unthinkably realistic proof was provided by 16 corpses that the SD & SS planted at the three scenes—10 dressed as German frontier guards and 5 in Polish Army uniforms at the buildings, plus 1 civilian at the radio station. The cadavers had been provided by a Silesian male civilian—the lone "casualty" at the radio station—that the Germans had wished to kill anyway and 15 male prisoners from the Sachsenhausen Concentration Camp. The latter were volunteers who had been told that, if they cooperated in these operations, they would be given their freedom. After leaving camp under Gestapo guard, their promised "freedom" came prematurely at the "action sites"—most provided by a Gestapo physician in the form of lethal injections. Then, to conform to the script (cynically called Operation CANNED GOODS), which required battle casualties, their bodies were shot.

This was a plausible scenario. But It failed operationally on the night of August 31st through multiple blunders by SS Major Alfred Naujocks and his covert team at the Gleiwitz radio station. However, because Naujocks' radio technician did not understand how the real Gleiwitz station's controls worked, he was unable to connect to the powerful radio transmitter at Radio Breslau. Consequently, instead of feeding into worldwide radio networks, the forged message only went out on the local frequency with a low-powered range of only some 25 miles. Finally (like the Watergate bunglers in later years) they fled the scene of their crime in such haste that they forgot to take along incriminating evidence—in this case their portable radio set for communicating with Berlin. By having abandoned it and their murdered civilian *outside* the station, they undermined their claim that the Polish raiders had actually taken over the station and its broadcasting equipment.

This inept handling of these incidents completely undermined their intended effect. The foreign press paid no notice. Even Heydrich's subsequently planted story in the Nazi Party's official newspaper, the *Völkischer Beobachter*, with its lame excuse that attributed broadcast failure to "the aggressors [having] cut the relay lines to Breslau"—failed to convince the target audiences. Surprisingly, Hitler was pleased so the bunglers weren't punished. However, the long-range consequence was a serious backfire. When the facts soon emerged during the subsequent war, the operation's monstrous brutality (much less the "Keystone Kops" buffoonery) became effective Allied propaganda against the Nazi enemy.[21]

CASE 1.4.2.2:
German Radio Deception ELEPHANTIASIS Fails to Deter a Russian Attack, 1942

Operation ELEPHANTIASIS was a German radio deception carried out by Lieutenant Fritz Neeb in early 1942. His transmissions were intended to deter the Russian forces from attacking a weakly held section of the Eastern Front near Vyazma by pretending it was heavily defended. The deception failed completely—the Russian did attack in overwhelming force.

Perhaps the Soviet Intelligence had actually detected the deception. However, the most likely explanation is that because it was only carried out through a single channel ("single-source") the Soviets simple rejected it as implausible and went ahead with their attack.[22]

---

21 André Brissaud, *The Nazi Secret Service* (London: The Bodley Head, 1974), 214-237; Deception Research Program (1982), 6-10, which summarizes Brissaud; and *Wikipedia*, "Gleiwitz incident" & "Operation Himmler" (both accessed 21 Mar 2010).

22 Maxim (1982), 22-24, based on the author's interview with David Kahn who, in turn, had interviewed the German radio intelligence officer, Fritz Neeb, on 30 Dec 1972. For Neeb see David Katz, *Hitler's Spies* (New York: Macmillan, 1978), 200-201. *FM 90-2* (1988), Chapter 7,

CASE 1.4.2.3:
Dakar: British Deception Fails, Sep 1940

Following the German occupation of France, the British and De Gaulle's Free French planned to recapture the strategic French naval base at Dakar from the pro-Nazi Vichy French regime. This ill-named Operation MENACE is particularly interesting because it is one of the few cases where strategic deception was used and failed, the attacker utterly unable to achieve surprise.

The British deception plan was simply too thin, consisting only of a weak effort to make Aden seem the port for which the large amphibious expedition was headed. Then, having lost the element of strategic surprise, the expeditionary force also sacrificed tactical surprise by issuing an ultimatum upon its arrival off Dakar. After two days of unsuccessful aerial and naval bombardment and an aborted landing of troops, the British and Free French slunk away.[23]

1.4.3. Target Judges It Irrelevant or inconsequential

The Problem:

The target become aware of the effect but doesn't see how it might be important enough to take any action.

The Solution:

Explicitly & intensively threaten some higher-value target.

CASE 1.4.3.1:
Hitler's Pre-invasion of Britain Propaganda Fails, 1940

Hitler opened a large-scale propaganda barrage against Britain in early 1940. Run from February to October, it was designed to demoralize British morale. Ideally provoking actual surrender or, at least, softening that target for Operation SEA LION, for his planned invasion of the British Isles. The propaganda claims of German power, capabilities, and resolve were not only accepted by the main target—the British public—it was even accepted by British Intelligence. However, it failed in its purpose. The British public and government held firm.[24]

---

copies Maxim (1982).
23 Whaley, *Stratagem* (1969/2007), Case A24 ("Dakar"), p.270 (in 2007 reprint).
24 Peter Fleming, *Operation Sea Lion* (New York: Simon and Schuster, 1957), particularly 117-119, 128n; Maxim (1982), 12-17; FM 90-2 (1988), Chapter 7, copied from Maxim.

CASE 1.4.3.2:
IRONSIDE Fails to Divert German Attention to Bordeaux, 1944

Operation IRONSIDE was the small-scale Allied deception operation in the weeks immediately before & after the Allied D-Day landings in Normandy in the English Channel and the southern French Mediterranean coast. It was designed to divert German reinforcements to the French Atlantic coast—specifically Bordeaux on the Bay of Biscay.

Conducted solely by the British Double-Cross agents it was noticed by German intelligence but judged of too slight a threat to require any reinforcement of the local Bordeaux area garrison. In other words, by attempting to sell the deception story through only a single channel (the double agents) the operation lacked convincing depth; and, by threatening a relatively low-value target, it lacked interest.[25]

CASE 1.4.3.3:
FORTITUDE NORTH: The Allied Feint against Norway, 1944
An faked threat to German occupied Norway draws only a yawn.

Hoping to keep the 12-division strong German 20th Army tied down in Norway when the massed Anglo-American forces hit the D-Day beaches in France, the Allied deceivers in London mounted a fairly large scale deception operation. Because it was plausibly staged in Scotland with the purely notional British Fourth Army, the deceivers could not rely on German reconnaissance flights from distant Norwegian bases to spot visual deception activities at Scottish bases or ports. Consequently only two channels for deception were used—double agents and faked radio traffic.

The Germans accepted the reality of the British Fourth Army and its threat to Norway. However, Hitler was more concerned to protect his Scandinavian supplies of strategic minerals from the *Russian* threat. Consequently, 20th Army's radio intercept efforts were tuned not to the British military traffic but to the Russian.

The British knew from their ULTRA reading of the German military traffic that all German units in Norway were being held in place. Ironically, they took this as conclusive evidence that their deception had worked when, in fact, Hitler would not have moved those units regardless of any British plan. As Dr. L. Daniel Maxim remarked, this was a case where "evidence consistent with one hypothesis may also be consistent with other views."[26]

---

25 Holt (2004), index; Cruickshank (1979), 159-160; Maxim (1982), 37-42; and *FM 90-2* (1988), copying Maxim.
26 Maxim (1982), "Maxim 8. The Importance of Feedback", citing Kahn and Lewin; Holt (2004), 559.

1.4.4. Target Finds the Message Ambiguous

The Problem:

The target notices the offered clues but interprets them as fitting equally with both the false hypothesis you intend and some other but irrelevant or even undesirable hypothesis.

The Solution:

Clarify—sharpen the focus of your message.

This problem exists to some degree in most deceptions, at least in their early stages. Usually that is OK. Indeed, you want your target to have to work out the connection between the deceptive clues and your intended hypothesis. Otherwise the target is apt to reject the clues as too suspiciously good to be true. Obvious clues will work only with the most gullible or naive targets.

1.4.5. Target Misunderstands the Intended Meaning

The Problem:

The target notices the offered clue but misconstrues their meaning.

The Solution:

Reframe the message by putting it in different words, visuals, or sounds—ones known to fit the collection requirements of the opponent's Intelligence Service.

Although I am unaware of any clear historical examples, this situation where the deception is simply misunderstood is an obvious possibility. For example, Target A might misinterpret a deception campaign designed to convey bluff pressure against Target B as pressure on itself—and consequently take preemptive action that the deceiver had not intended.

1.5. Target Detects the Deception

> Deception can itself be dangerous if it is performed so ineptly that it provides the enemy with clues as to what are the real intentions.
>
> — R. V. Jones, "Intelligence and Deception" (1981), 20

The Problem:
Your target detects your attempt to deceive.

The Solution:
The Theory of Outs (as explained in Chapter 4.3).

The three previous categories pitted more-or-less incompetent deceivers against usually dull-witted targets. Now is the stage where we enter that proverbial battle of wits—one where a generally competent deceiver meets a generally competent target. Here the deceiver must avoid making plans that are 1) so complicated that they self-destruct, 2) so simple that they virtually red-flag themselves, or) that are vulnerable to the specific deception analysis techniques being used by the opponent.

1.5.1. Too Complicated: The Tangled Web Principle

> O, what a tangled web we weave,
> When first we practise to deceive!
> — Sir Walter Scott, *Marmion* (1808),
>   Canto 6, Stanza 17

> But, if we practise for a while,
> We'll get a wizard's knack for guile.
> — BW, 11 Jan 2010, with apologies to
>   W. Scott and J. R. Pope

Deception need not be complicated, much less complex. In fact, at its most effective, it is always a simple process. As that master British deceiver and counterdeceptionist, Dr. R. V. Jones, explained about Scott's tangled web:[27]

> I learned those lines at the age of thirteen from a schoolmaster who stressed the significance of 'first,' with its intriguing implication that practice might make perfect.

Indeed. It is a dysfunctional myth that deception operations must be complicated. Jones argued persuasively on this point by showing that the minimalist principle of Occam's Razor is usually (but not always) preferable to the clutter of "Crabtree's Bludgeon".[28]

---

27 Jones, *Reflections on Intelligence* (1989), 106.
28 Jones (1989), 87-88. A more rigorous argument for the basic simplicity of deception is Whaley (1982).

The Problem:
A major cause of deception failure occurs when it is made too complicated.

The Solution:
Dr. R. V. Jones not only had a deep understanding of how deception works but an even rarer knack for explaining it. His sentence quoted just above is an example. That simple 25-word statement not only frames the problem but gives the answer. Namely, we can learn from experience which seemingly necessary complications are, in fact, not needed.

CASE 1.5.1.1:
Operation MINCEMEAT, Spain 1943

> "Keep it simple, stupid"—KISS—is our constant reminder.
>
> — Kelly Johnson, lead engineer of Lockheed's "Skunk Works", *More Than My Share of It All* (1985), 161[29]

The famous "Man Who Never Was" ruse was, of course, a success—although less grand than usually portrayed.[30] And far more risky than recognized—indeed a very high risk proposition that invited failure and even backfire. It's execution had at least five potentially fatal flaws, glaring incongruities, virtual "red flags", that had German Military Intelligence (the Abwehr) noticed would have led to it's unmasking as a deception operation and, far worse, strong suspicion of the other interlocking deceptions.

The idea behind this ruse was to let German Military Intelligence (the Abwehr) find a batch of secret papers that taken together would suggest that the next Allied offensive would be invasions of Greece and Sardenia in the central Mediterranean—rather than the real target, which was Sicily. However, the planners, RAF Flight Lieutenant Cholmondeley (of B1a, which handled double agents) & Navy Lieut.- Commander Montagu (of LCS which coordinated deception), were not content with a set of faked documents, one that was almost foolproof. They push their luck by throwing in a dead human courier. That was a cute touch but one that, as with any impostor, required a mass of consistent details to sustain the impersonation.

---

29 This famous line is usually attributed to Kelly Johnson; although, in this his only authentic citation, he seems to attribute it collectively to his R&D group at Lockheed. However, it was a standard U.S. military in the Vietnam War by 1963 according to J.E. Leighter (*editor*), *Random House Historical Dictionary of American Slang*, Vol.2 (1997), 364. So, which came first?

30 Ben Macintyre, *Operation Mincemeat* (London: Bloomsbury, 2010); Holt (2004), 369-378; and Andrew (2009), 286-287.

First there was the matter of the theater tickets. The corpse of "William Martin" floated ashore in battle-dress of a British Royal Marine major, "secret" papers filling an attaché case chained to his waist and his pockets filled with personal items. Most of these papers and items had been fabricated with care. But one had a careless slipup—the theater-ticket stubs were dated a day later than would have been consistent with his travel orders. A small clue, perhaps, but one that the Abwehr analysts fortunately overlooked.

Second, the fingerprints of the corpse did not appear on any of the objects that where among his supposedly personal items. Yet all these objects had been handled by other persons, including Montagu, Cholmondeley, and at least one female B1a (Double Agents) secretary who posed as "Major Martin"'s fiancée, "Pam".

Third, and potentially much more serious, was the corpse's manner of death. The cause was pneumonia, a fact that could be readily proved by autopsy. But the pretend Major Martin's cover-story required that the cause was death by drowning, a falsehood that an autopsy could equally easily disprove. The British planners had made the potentially fatal error of assuming the remote Abwehr post in southeast Spain would not have a suitably trained physician available to examine the body. In fact, when the body was recovered, one of the top German forensic surgeons—an Abwehr man himself—was on the spot. Luckily for the British deception planners, the surgeon's colleagues simply did not even bother to consult him, relying only on the faulty diagnosis of the local Spanish physician who certified the cause of death as drowning.

Fourth, and a notably eyebrow-raising incongruity, a very bright red flag, was the condition of the corpse. British efforts at refrigeration had been inadequate—the body when found in Spain was already in a stage of advanced decomposition grossly inconsistent with its alleged recent death.

MINCEMEAT was also flawed at a higher level of planning where it tied into larger deception plans. In their self-congratulatory writings the MINCEMEAT participants and historians overlook the potentially disastrous consequences of its detection by the enemy. Had the Abwehr detected the deception, it would have implied that Sicily was the real target, which in turn would risk discrediting several of "A" Force's carefully built up deception assets.[31]

Through most of the 1970s I suspected the real reason British officialdom authorized Montagu's semi-official history of MINCEMEAT was that they recognized it was a flawed model of how to deceive. In other words, they cunningly published it to teach outsiders the false lesson that good deception plans were complicated plans. I am now satisfied that they were quite unaware

---

31 Mure (1980), 207, 210.

of this potential benefit and sincerely deluded themselves that MINCEMEAT was a prime example of excellent deception.

1.5.2. Too Simple: The Too Perfect Principle

This is the profoundly counterintuitive principle that a deception can be so perfect, so reduced to the minimum number (two) of incongruities that one and only one solution is possible—so obvious that the target audience is apt to detect the deception. This key principle of deception is the exclusive contribution of a few conjurors and one mystery story writer. First identified as a problem by American comedy magician Donald "Monk" Watson in 1945 and Canadian sleight-of-hand master Dai Vernon by 1992, it was codified in 1970 by American amateur close-up artist Rick Johnsson.[32] And most recently, in 2005, American amateur magus and law professor Christopher Hanna has applied it to tax accounting in a model that intelligence and deception analysts might profitably heed.[33]

Meanwhile, American expatriate mystery story writer Raymond Chandler had independently rounded on the same problem in 1948; and, like Rick Johnsson, offering the same solution, although framed in different words: [34]

> The most effective way to conceal a simple mystery is behind another mystery. This is literary legerdemain. You do not fool the reader by hiding clues or faking character à la [Agatha] Christie but by making him solve the wrong problem.

Rick Johnsson proposed one of the more original and stimulating theories in conjuring, specifically relating to the detection of magical deceptions. He stated (p.248) this theory as:

> "SOME TRICKS, BY VIRTUE OF THEIR PERFECTION, ARE IMPERFECT. SOME TRICKS, BY VIRTUE OF THEIR IMPERFECTION, ARE PERFECT"

Johnsson explains this is by detailed examples and then concludes (p.250):

> It behooves magicians to avoid leaving a spectator one accurate path to follow, leading to the modus operandi; or to leave

---

32　Rick Johnsson, "The 'Too Perfect' Theory," *Heirophant*, No.5/6 (Fall-Spring 1970/71), 247-250.
33　Christopher H. Hanna, "From Gregory to Enron: The Too Perfect Theory and Tax Law," *Virginia Tax Review*, Vol.24, No.4 (Spring 2005), 737-796.
34　Raymond Chandler, "Twelve Notes on the Detective Story, Addenda" (revised 18 Apr 1948), as published in Frank McShane (*editor*), *The Notebooks of Raymond Chandler* (NY: The Ecco Press, 1976), 38. Chandler paraphrased himself the following year in his manuscript notes: "[T]he only reasonably honest and effective method of fooling the reader ... is to make the reader exercize his mind about the wrong problem ... which will land him in a bypath because it is tangential to the central problem. And even this takes a bit of cheating here and there." See D. Gardiner & K. S. Walker, *Raymond Chandler Speaking* (Boston: Houghton Mifflin, 1962), particularly pp.68-69.

the onlooker paths that take credit away from the magician himself. It is better to direct the spectator to follow a path of the magician's own choosing, leading him to the conclusion that the magician is "some clever devil". That's the name of the game, baby."

I'd condense & generalize this as:

The fewer incongruities in a deception, the more likely the victim will detect it by process of elimination. Therefore, add just enough incongruities (false clues) to misdirect the victim away from all obvious solutions. (Whaley, 26 Jan 2010)

On this point see also Adam Gopnik (2008) in writing:

> Of all the arguments that can preoccupy the mindful magician, the most important involves what is called the Too Perfect theory. Jamy Ian Swiss has written about it often. Presaged by Vernon himself, and formalized by the illusionist [slight-of-hand artist] Rick Johnsson in a 1971 [1970] article, the Too Perfect theory says, basically, that any trick that simply astounds will give itself away. If, for instance, a magician smokes a cigarette and then makes it pass through an ordinary quarter, the only reasonable explanation is that it isn't an ordinary quarter; the spectator will immediately know that it's a trick quarter, with a hinge. (Swiss wrote that once, after he performed the Cigarette Through Quarter—perfectly, in his opinion—a spectator responded, "Neat. Where's that nifty coin with the hole in it?")
>
> What makes a trick work is not the inherent astoundingness of its effect but the magician's ability to suggest any number of possible explanations, none of them conclusive, and none of them quite obvious. For an application of this theory to deceptive practices in tax law see Hanna (2005).

The Problem:

A deception can be so simplistic that its solution is almost obvious.

The Solution:

Strong misdirection—simple but strong enough to lead the target onto a false trail leading to a satisfying but wrong solution. One example was the way Hitler, having decided to abandon his planned invasion of Britain in 1940

(Operation SEALION), turned that threat into the cover story for his new plan for a surprise invasion of Russia (Operation BARBAROSSA).[35] Three additional cases are:

CASE 1.5.2.1:
Straw Soldiers, China AD 755

> The Italians have a Proverb, He that deceives me Once, it's his Fault, but Twice, it is my fault.
>
> — Sir Anthony Weldon, *The Court and Character of King James* (1650)[36]

The idea of switching a real person or object for a dummy or vice versa is what magicians call the Double-Exchange Principle. Although relatively rare in Western military culture, it is quite common in China. There it was codified in the 36 Stratagems as Stratagem No.7, "*Wu Zhong Sheng You*", which I translate as "Make the Illusion Real" or "From Nothing Make Something". This 7th principle was probably inspired by Taoist philosopher Lao Tzu's somewhat counterintuitive notion that an illusion can conceal a reality. Compiled anonymously sometime around 1644, this is one of the best known and widely read of all Chinese texts on strategy—military and otherwise.[37]

One simple illustration of the Double-Exchange Principle is the famous Chinese story of "Straw Dummies for Soldiers".[38] But it also illustrates the Too Simple Principle:

In AD 755, the illustrious T'ang Dynasty was threatened by a major rebellion. One rebel general, Ling Huchao, set siege to the city of Yongqiu, which was defended by a small garrison commanded by loyalist Commander Zhang Xun. This wily officer commanded his soldiers to make a 1,000 man-sized dummies of straw dressed in black clothing, attach them to lines, and let them slide down the outside of the city walls at twilight. Rebel General Ling, misperceiving this as a sortie by the city garrison ordered his archers to loose a hail of arrows against them. When Commander Zhang had his straw dummies drawn up with their thousands of "captured" arrows, rebel General Ling realized he'd been tricked.

---

35   Whaley, *Codeword BARBAROSSA* (Cambridge, Mass.: The MIT Press, 1973), 172-174 where this technique of hiding a new deception within an old one was first explicitly identified.
36   This is the earliest source I find for the popular modern proverb, "Fool me once, shame on you; fool me twice, shame on me."
37   A particularly useful analysis of Stratagem #7 is Harro von Senger, *The Book of Stratagems: Tactics for Triumph and Survival* (New York: Viking, 1991), 85-108.
38   Von Senger (1991), 88-89.

Next evening, Commander Zhang had 500 real soldiers lowered down the city walls. General Ling, thinking them dummies sent to harvest more arrows, laughed in derision and made no preparations for battle. Zhang's small force struck hard & fast, set fire to the Ling's camp, killed many rebels, and scattered the rest like straw in the four winds.

This is a rousing legend and it is probably based on fact. But, if so, not all the facts, unless we assume that rebel General Ling was abysmally slow witted. A single repetition of any such ruse is seldom enough to induce the Cry-Wolf Effect. Any sensible commander would at least have gone on high alert.

CASE: 1.5.2.2:
Russian Tactical Radio Deception, Eastern Front WW2

The Too Simple Principle probably explains the general lack of effectiveness of most of the tactical radio deception operations played by the Russians on the Eastern Front in WW2 against the Germans. So concluded Wehrmacht Lieutenant Fritz Neeb who headed the German Army's radio section on the Central Front. Neeb was intercepting about two of these sets of games each week. After the first few, this was too frequent and too long to be part of a cry-wolf campaign to lull the Germans. Moreover, because the radio deception was not corroborated by any other means through any other channels, this was the least convincing type of deception—single channel ("single source"). In any case they became almost self-revealing deceptions, ones that readily yielded to even the simplest deception analysis.[39]

CASE 1.5.2.3:
Allied ACCUMULATOR Feint toward the Cotentin Side of Normandy

To induce the defending German units on the Cotentin Peninsula to stay in place and not move to reinforce the real Allied D-Day landings on the Normandy beaches some 50 miles eastward, a simulated diversionary attack was directed at the Cotentin on D+6 & 7. Hastily patched together by Col. David Strangeway's "R" Force, this operation employed only two radio sets broadcasting in plain (uncoded) voice cruising toward the peninsula on two Canadian destroyers, one of which soon aborted through engine trouble. Thus this was, in effect, a single channel ("single source") deception with no confirming depth much less breadth—the perfect formula for any deception failure.

---

39 Maxim (1982), 24-25, based on the author's personal communication with David Kahn and the latter's 1972 interview with Neeb. Copied in *FM 90-2* (1988), Chapter 7.

This deception operation was of such small scale, inept execution, and zero consequence that I know of only three published accounts have bothered to discuss it. Otherwise it is a mere footnote.[40]

1.5.3. Target Succeeds at Deception Analysis
The previous section examined cases where deception operations were so obvious—lacking both depth & breadth—as to be virtually self-disclosing. This section looks at the situation where serious deception analysis is required to detect sophisticated deception operations.

The Problem:
Even when faced with a competently designed deception, the target always has the possibility, through systematic analysis of the incongruities present, of fathoming the deception.

The Solution:
Eliminate any clues or incongruities that are apt to immediately direct the target's attention to the one correct solution. A fine example follows.

CASE 1.5.3.1:
Jones and the Telltale Decoys, German Occupied France 1944
Because Dr. R. V. Jones, as Chief Scientist of the British Air Ministry, was also his own deception analyst he was able to unmask a German deception in WW2 due to its thoughtless retention in the decoy of a clue that helped Jones identify the nature of Hitler's latest secret weapon. These were the powerful V2 terror rockets that were about to start bombarding London. This story is best told in his words:[41]

> "As with all problems involving security, deception can itself be dangerous if it is performed so ineptly that it provides the enemy with clues as to what are the real intentions. In the deployment of the V2 sites in France for the bombardment of London, for example, I was able to deduce the intended rate of firing because of the decoy sites which the Germans had erected in the hope that we would bomb them instead of the actual storage sites. To make sure that we spotted them, the Germans

---

40 Cruickshank (1979), 200-201; Strangeways in M. Young & R. Stamp (*editors*), *Trojan Horses* (London: Bodley Head, 1989), 46; Maxim (1982), 34-37; *FM 90-2* (1988), Chapter 7, coping Maxim. Plus passing mention in Holt (2004), 807.

41 Reginald V. Jones, "Intelligence and Deception," in Robert L. Pfaltzgraff, Jr & Uri Ra'anan (*editors*), *Intelligence Policy and National Security* (London: Macmillan, 1981), 20-21. Further details are in R.V. Jones, *The Wizard War* (New York: Coward-McCann & Geoghan, 1978), 453-454.

made the decoy sites obvious, but they forget to maintain any activity and so there were no paths or wearing of the grass as there would have been if the sites were being worked.

There were twenty decoys, deployed roughly from Calais to Cherbourg, with fourteen east and six west of the Seine. When we landed in Normandy we, of course, captured not only the decoys west of the Seine, but also the real storage sites, and could easily establish their total [rocket] storage. Then, assuming that there was the same proportion of genuine storage sites to decoys east of the Seine, this gave the total storage sites as about 400; since it appeared to be German practice to keep two weeks' supplied forward, this suggested an intended monthly rate of fire of 800 [rockets], which was a fairly close estimate to the actual rate of fire intended, 900 per month.

I could not have made such a close estimate had the Germans not tried to deceive us."

CASE 1.5.3.2:
Bay of Pigs: US Deception Seen Through, Apr 1961

Cuba's Fidel Castro and his intelligence service had ample warnings of the planned US Bay of Pigs operation. They received these through their own sources, Soviet diplomatic and KGB channels, and rather blatant comment in the US press. Having thereby a clear picture of the ground truth, they readily saw through the rather weakly designed and executed US deception operations.[42]

---

42 Whaley, *Stratagem* (1969/2007), Case A65 ("Bay of Pigs"), pp.486-502 (in 2007 reprint); and *Wikipedia*, "Bay of Pigs Invasion" (accessed 12 Jul 2010).

2. Degrees & Types of Failure

Nearly all military writers and researchers oversimplify the issue of deception failure by their tendency to treat it as an either/or matter. Both failure and success are, at best, only rough characterizations of the outcomes of any deception operation. Deception is almost never either fully successful or a total failure. In reality, deception (like surprise) is almost always only more-or-less successful. Indeed, deception (again like surprise) can vary greatly in both intensity and variety.[43]

To elaborate, the success or failure of each deception operation can best be defined by the deceiver's own expectations. Some expectations will be subjective—is the target surprised, confused, demoralized, etc. Other expectations are measurable in terms of amount of geographical ground taken, properties seized, targets captured, casualties (or casualty ratios) inflicted, etc. In other words, the results of deception should be measured along a continuous scale on which total "success" and "failure" fall toward the extremes. And I stress that neither "success" nor "failure" mark the extreme ends of this success-failure continuum because, as found in another study, outcomes are often far above expectations as well as occasionally ending in blowback. Moreover, both success and failure can have important long-range consequences (which may be positive or negative. For example, deception failures, particularly repeated ones, will often lead to severe degradation of the unmasked deceiver's general credibility. Even an immediate "success" can lead to long-range dysfunction, as when high enemy casualties can inspire increased enemy recruitment. The following figure illustrates this range of outcomes.

Figure: The Deception Success-Failure Continuum

LRC—EXCEEDS—SUCCESS/FAILURE—BACKFIRES—LRC
(Where **LRC** means long-range consequences.)

Deception planners and analysts should be acutely aware of this continuum of potential outcomes—particularly the potential for both a success that exceeds their expectations as well as one that backfires. However, both extreme potential outcomes (the long-range consequences) fall outside their area of competence and responsibility and are best left to the assessment of higher levels of military or political command. Indeed, vetoes by high commands are one major cause of deception aborts. (See Chapter 2.5).

---

[43] As first described in Whaley, "The Varieties and Intensities of Surprise and Deception" in *Stratagem* (1969/2007), Chapter 5, pp.111-118 in the 2007 reprint).

We can now turn from the degrees of deception failure to the types of things to which it can apply:

Misperception in general and deception in particular apply to only nine types of things. These are the things that deceivers try to dissimulate or simulate, hide or show:[44]

1. CHANNEL — The media through which the deceiver communicates.
2. PATTERN — The structure and process of the deception (defensive, offensive, etc).
3. PLAYERS — The actors (the enemies, allies, neutrals, etc).
4. INTENTION — What the deceiver hopes to accomplish.
5. PAYOFF — The deceiver's perceived costs and benefits.
6. PLACE — Where the event will occur.
7. TIME — When it is scheduled for.
8. STRENGTH — The amount of personnel & material the deceiver will bring to bear.
9. STYLE — New methods or technologies to be introduced.

For every operation, deception can be attempted in any one, several, or all of the above categories. And each can be successful, partly successful, or a total failure. Moreover, as with British General Alexander at Massicault (Case 2.1.2), the Commander can fake threats at more than one place, at least one of which may succeed.

There are pronounced differences in the frequency with which each category occurs in military operations. My survey of 108 modern battles involving surprise at the strategic and large-scale tactical levels found the following variation:

---

44 Based on Bart Whaley, "A Typology of Misperception; or The Ways We Can Be Wrong", draft March 1980, 111pp. Some of this material was published in J Barton Bowyer [pseudonym of Bart Whaley and J. Bowyer Bell], Cheating (New York: St. Martin's Press, 1982), 428-431, where the category of "pattern" was added. My most recent exposition is in Whaley, *Deception Verification in Depth & Across Topics* (FDDC, Nov 2009), 30.

Table: Frequency of the Nine Types of Surprise in Military Operations, 1914-1968

TYPES OF SURPRISE	NUMBER	%
Place	78	72.2
Time	71	65.7
Strength	62	57.4
Intention	36	33.3
Style	28	25.9
Payoff	3	2.8
Players	2	2.8
Channel	1	0.9
Pattern	1	0.9
N=	(108)	

Further analysis showed that, while all nine categories applied at the "strategic" level of operations, "tactical" operations tended to be equally good at manipulating only the PLACE, TIME, STRENGTH, and STYLE categories. Tactical ops were significantly weaker at manipulating INTENTION and simply didn't or couldn't manipulate PAYOFF, PLAYERS, CHANNEL, or PATTERN.[45] The superficial reason is often a simple lack of appropriate deception assets at the tactical level, but the underlying reason is that tactical operations typically develop too quickly to permit the buildup of these latter modes of deception.

Although these are the frequencies in the historical record, they need not hold for the future. I predict that, if military deception planning grows more sophisticated, we shall begin to see systematic efforts devoted to deception about the hitherto more neglected categories and at both the strategic and tactical levels. There are hints that the Russians have already been doing this for the category of "style".

The four types of failure are Partial, Total, Backfire, Turnabout, and Aborts. Additionally, two different types of failure are discussed—Pseudo and Potential Failures. But, let's begin with the Partials.

---

45 On these 9 categories and their evolution see Whaley, *Stratagem* (1969), 214-215, 248-249; and Whaley, *Textbook of Political-Military Counterdeception* (FDDC, Aug 2007), 25-31.

2.1. Partial Failure

Even when deception fails, it is almost always only a partial failure. Four cases in point, beginning with a most instructive example of a deception plan and operation that back-fired, producing an effect opposite of the main (strategic) goal intended by the deceiver, yet still preserved the secondary (tactical) goal.

CASE 2.1.1:
Col. Clarke, Italian East Africa 1941

When Colonel Dudley Clarke joined General Wavell in Cairo at the end of 1940 to found and lead the so-called "A" Force deception team, his first operation was to devise the strategic and tactical cover plans for the forthcoming invasion of Italian East Africa.

Clarke's strategic plan was designed to make the Italians think he was about to attack them from the south in order to induce them to reinforce there, weakening their northern flank where Wavell's real attack would come. The problem was that this strategic deception "succeeded" in the sense that the enemy "bought" it. But the result was just the opposite of what Wavell wanted. Deceived into expecting the main attack in the south, the Italians abandoned southern Ethiopia and reinforced the north.

The planner must decide how he wants his target to *react* in a given situation. The lesson learned was, as Colonel Clarke admitted: "of inestimable value."[46]

> After that it became a creed in "A" Force to ask a General, "What do you want the enemy to *do*?", and never, "What do you want him to *think*?"

Clarke's discovery points to and clears up the ambiguity in the words "success" and "failure" as applied to deception. If we apply them to the effect on the target's thoughts and beliefs, we land ourselves in the logical paradox met in the previous paragraph. To eliminate this paradox, success or failure should be measured only by the target's actions.

This, incidentally, is a very rare example of a deception operation that not only failed but backfired. However, while the *strategic* deception failed, its *tactical* backup didn't; and the Italians were surprised by the precise details of the attack in the north. Moreover, Clarke immediately applied the lesson of his

---

46 Brigadier Dudley Clarke, "Some Personal Reflections on the Practice of Deception in the Mediterranean from 1941 to 1945" (memo dated 6 September 1972), as published in David Mure, *Master of Deception* (London: Kimber, 1980), 274. See also Mure (1980), 9, 81-82; Holt (2004), index; and Daniel & Herbig (1980), 4.
 Curiously, the official history of deception, Howard (1990), 34-35, rates CAMILLA a "complete surprise", overlooking Clarke's rule that it's not what you want the opponent to think but to do. By Clarke's criteria, CAMILLA succeeded in getting the Italians to think right but do wrong.

initial strategic failure: henceforward during the East African campaign he arranged for the Italians to discover the exact place of each British attack—a stratagem that insured they would meet no opposition.[47]

The final result was that after only three months of campaigning, Wavell's small force had seized the half-million square miles of Italian East Africa (Ethiopia, Eritrea, and Italian Somaliland), reconquered British Somaliland, and obtained the surrender of the entire 250,000-man Italian Army at a tolerable cost of about 4,000 British Empire casualties.

CASE 2.1.2:
Gen. Alexander at Massicault, Tunisia 1943

The official U.S. military historian for the Northwest African campaign concluded that General Alexander's deception plan for the final battle in Tunisia failed because the enemy Commander "expected the main attack by First Army in the general area where it was launched."[48] This historian got the facts right but jumped to a wrong conclusion. Let's examine this claim.

For his breakthrough to Tunis in 1943 that evicted the Axis from North Africa, General Sir Harold Alexander adapted an elaborate deception plan to convince Colonel-General Jürgen von Arnim—Rommel's replacement—that his next offensive would came from Montgomery's stalled Eighth Army on the right flank, while Lieutenant-General Kenneth Anderson's First Army would make the real attack through the center to Massicault and thence down the Medjerda Valley. Alexander's deception plan was designed to create "a delicate balance of forces", specifically a calculated imbalancing of the enemy.[49]

In fact, Arnim expected the offensive by First Army rather than Eighth. This correct assessment was due to poor Allied radio security, intercepted Allied radio traffic having disclosed the shift of two divisions from Eighth to First. Henceforward Arnim largely ignored the bogus threat from Eighth and focused on First. Thus Alexander's *strategic* deception plan had indeed failed—at least in its main purpose.

However, the American official historian missed the subtlety of Alexander's deception plan. First, Alexander's stratagem had sown enough uncertainty and misperception with Arnim that when the blow fell the German line was still badly imbalanced, reinforcements to the "expected" general area consisting of only one Panzer regiment and a single anti-tank battalion. Second, Alexander

---

47 Mure (1980), 81-82.
48 George F. Howe, *Northwest Africa: Seizing the Initiative in the West* (Washington, DC: Office of the Chief of Military History, 1957), 647-648.
49 For details of the situation, British plans, intelligence, & deception, and the German intelligence appreciation see Whaley, *Stratagem* (1969/2007), Example B30 (Massicault).

had, like Wavell and Clarke in the previous example, imbedded tactical deception within his strategic one.

Indeed, Alexander had used two tactical deceptions, both successfully. The first was a strong diversionary attack mounted on D-minus-15 by British IX Corps at the extreme left of center. That 5-day operation succeeded in provoking the enemy to overrespond by diverting much of his armor and committing almost all his mobile reserve to that sideshow. Alexander's second tactical deception used a double-agent radio-game on D-minus-3, plus visual signatures that Luftwaffe reconnaissance dutifully reported, as well as some other means. All these were orchestrated to sell the notion that one British division was about to redeploy from the center to the extreme right of center to penetrate the German front at that point. The result was that the German Command immediately switched a major armored unit from the vital center to the irrelevant south.

The subsequent break-through battle (and follow-up) was a stunning success for Alexander. By D+6 the entire Axis army of 248,000 had surrendered, only 663 escaping to Italy. Allied casualties were 2,000.

CASE 2.1.3:
The Errant Airborne, Normandy 1944

Historians, commanders, and soldiers who evaluated the effectiveness of parachute combat drops before the late 1900s agree on one point. The single most important factor working against the effectiveness of paratroop operations had been the failure to drop where planned. Moreover, doing so in a wide scattering of units with the consequent time consumed in regrouping and moving cut to the more-or-less unintendedly distant objective. These early writers believed these delays cost the element of surprise on which most combat drops depended and in which deception was almost always an adjunct. These early self-critics were unnecessarily harsh on themselves; and their conclusion wrong because they failed to look on the "other side of the hill".

The deception plans and operations in support of early paradrops also failed to the extent that they landed nearer-or-further from the intended drop-zone. The planners found this situation as frustrating as the paratroops. Again, like the paras, they missed the same important point by failing to look at their operations from the perspective of the enemy. What did, in fact, happen? Let's look at the three-division airborne operation in support of the Normandy D-Day landings.

On D-day, 6 June 1944, three Allied airborne divisions plopped onto Normandy, well inland from the assault beaches. Their mission was to provide a blocking force that would slow the approach of German reinforcements until the beachhead was secured. The operation was covered by diversionary drops of dummy paratroop units.

The actual drops were badly scattered. So much so that they often overlapped, thereby negating the dummy drop zones. Worse, it took many precious hours for the paratroops to reconstitute themselves as fighting divisions. Although this was the typical result of airborne operations before helicopters could pin-point deliver combat troops, it had not been anticipated by the SHAEF planners. Consequently the Allies considered this part of the invasion to be its least effective action, a virtual failure. But was it? What was the enemy reaction?

The Germans were immediately and fully aware of this wide dispersal of the airborne units. But they assumed it was deliberate, and placed dozens of battle-ready Allied regiments and battalions at key points on their Normandy battle map. To deal with these units, the Germans divided their own forces and sent them forward slowly and cautiously—thereby missing a grand opportunity to either destroy the lightly armed paras in detail before they could reorganize or simply ignore them and move directly against the beachhead.

Ironically, the unplanned scattering of the airdrop had as much or better effect than the planned version because it achieved its primary mission of slowing the enemy reaction, albeit inadvertently by confusing him rather than fighting him. Even the dummy drops added to the enemy's confusion.[50]

Empathy, the ability to think oneself into another person's mind, is a valuable quality in a Commander or his deception planners. In evaluating the degree of success or failure of our "best-laid schemes", we should always factor in the deception target's perceptions and reaction.

CASE 2.1.4:
The Score of 24th Hq Special Troops, Western Europe 1944-45

The U.S. Army's 24th Headquarters Special Troops generated a total of 26 tactical deception plans in support of General Omar Bradley's 29 U.S. divisions during the ten months from Normandy in June 1944 until March 1945. Of these 26 plans, we see from the table below, 5 aborted (for a variety of reasons discussed in Case 2.5.5) and 21 became actual operations. Of these latter, 4 were entirely successful, 11 were at least partly successful, and only 1 was known to have failed completely (Case 2.2.10). Of the 5 operations whose results are unknown, the very lack of evidence suggests possible failure (or only negligible success). On that most pessimistic assumption we get 15 "successes"

---

50 Such chaotic military para drops were common in WW2 and for a decade or two after. That is no longer true. By the late 1970s the choreography of mass airborne drops had been brought to a high level of coordination, as vividly described to me by former U.S. Army jump master, Col. (USA, Ret) Hy Rothstein, 2 Feb 2010.

as against 6 "failures" plus 5 aborts—a clearly cost-effective record given the small size of the unit (1,106 officers and men, mainly engineers and signals personnel) and the nature of the German reaction to its deceptions.[51]

Table: Results of Deception Plans by 23rd Hq Sp Tr, 1944-45

RESULT	No.
Success	4
Part Success	8
Probable Part Success	3
Failure	1
Unknown	5
Aborts	5
TOTAL PLANS:	**26**

2.2. Total Failure

That early and most pessimistic historian of deception, Charles Cruickshank, was extraordinarily wrong-headed in concluding that, "It may be a fair assessment that the deception planners [in WW II London] just about broke even before *Overlord*."[52] Cruickshank thoughtlessly confused the several mere failures to sell specific plans with outright backfire. In other words he was unaware that meaningful success/failure assessments can fall anywhere along a continuum whose opposite ends are those plans that drew the intended enemy response versus plans that backfired in such a way as to cause great harm to the intended deceiver. None of the cases he cites—both failures of salesmanship and aborts—caused actual damage. Cruickshank's either/or analysis is false but pernicious. I have personally observed Cruickshank being accepted and cited by one senior American deception research project director at the end of the 1970s when he completely discredited or questioned particular historical deception operation.

In fact, only rarely does deception fail completely. And even then seldom with serious consequences. Let's examine a few examples:

---

51 *Official History of the 23rd Headquarters Special Troops* (1945). The official history of this unit was hastily written immediately after the war and with little access to German records or interrogations and none to secret Allied Intelligence such as the feedback from ULTRA. We now know that two (and probably more) of the cases were far more effective than the historian reported. Thus the following table on Results of Deception Plans adjusts for these two later known outcomes.

52 Cruickshank (1979), 220.

CASE 2.2.1:
Operation CONDOR, French Indochina 1954

As the noose tightened around the besieged French force trapped at Dienbienphu (Case 2.4.2), General Henri Navarre, commander of the Army in French Indochina, was repeatedly urged by his staff officers to launch large-scale diversionary feints. Navarre rejected this advice and the several pin-prick diversions that he did mount failed to divert the Viet Minh commander, General Vo Nguyen Giap from his main prize.[53]

There is some slim evidence that the largest of these diversionary efforts, Operation CONDOR, did generate considerable—but insufficient—pressure within the Viet Minh leadership to persuade Giap to lift his siege of Dienbienphu. CONDOR was a four-battalion, 3,100-man operation. Its movement overland from Laos toward Dienbienphu in April attempted to simulate the vanguard of a (nonexistent) major offensive to relieve the embattled garrison. It did this by incorporating small detachments from French units throughout Indochina in the expectation that as some fell into the hands of enemy Intelligence, the Viet Minh might believe the ruse. The French decoys were themselves successfully deceived on this point, and it may have contributed to Giap's momentary second thoughts.[54]

CASE 2.2.2:
Operation PURPLE WHALES, Burma 1942

Colonel Peter Fleming headed General Wavell's small deception planning team in India. This was "D" Force, a direct spinoff of Dudley Clarke's "A" Force in Cairo.

Fleming's second deception plan (for his first see Case 2.2.3) was launched with Wavell's endorsement. Operation PURPLE WHALES was an ambitious scheme to sell Japanese Intelligence three strategic lies: 1) that a Second Front in Europe was imminent, 2) that the British were going on the defensive in the Middle East; and 3) that the Allied buildup in Southeast Asia had reached a point where they were about to go on the offensive against the Japanese. These false notions were embedded in a 13-page document represented as the minutes of a meeting of the Joint Military Council, which sat in Chungking. This document had been fabricated with the assistance of Chief of the Imperial General Staff Sir John Dill in London and approved by Chinese President Chiang K'ai-shek.[55]

---

53 Bernard B. Fall, *Hell in a Very Small Place: The Siege of Dien Bien Phu* (Philadelphia: Lippincott, 1967), 41-43.
54 Fall (1967), 228, 317-318, 342, and index under "Condor".
55 Duff Hart-Davis, *Peter Fleming: A Biography* (London: Jonathan Cape 1974), 271-274.

For his channel to enemy Intelligence, Fleming chose a Chinese agent who sold it to the Japanese for a convincingly sizeable sum. The scheme apparently died at that point. No evidence ever emerged that PURPLE WHALES had succeeded. Fleming later suspected that the text may not have been concocted with sufficient subtlety to persuade the Japanese Intelligence officers of its authenticity. Whatever the reason, no harm was done except for a few minutes of lost time by CIGS Dill and President Chiang.

CASE 2.2.3:
Col. Fleming's Haversack-type Plan Fails, Burma 1943

Operation FATHEAD was Peter Fleming's attempt to replicate Meinerthagen's famous (but miscredited & possibly failed) Haversack Ruse.[56] In this case a suitable corpse, fitted out with radio, code, and operating instructions, was airdropped behind the Japanese lines in Burma. Fleming's hope was that Japanese Intelligence would try to operate this set back, as if they had a double agent. However, the Japanese did not do so.

Having no feedback, Fleming could only speculate about why the Japanese had not taken the bait. Perhaps the corpse was never found. Perhaps his dropped radio had been damaged beyond repair. Perhaps local Japanese Intelligence felt they lacked the language skills needed to imitate a British agent. Perhaps they simply could not work out the codes.[57]

CASE 2.2.4:
Dissimulative Camouflage Fails, Germany 1943

In 1943 the German Navy suddenly began losing U-boats at an alarming rate. Their Naval Intelligence soon received intelligence that the British had developed an airborne infrared detector. For several months German scientists worked frantically to invent an anti-infrared paint and hastily recoated their U-boat fleet. This paint was superb; it would have effectively camouflaged the U-boats against infrared detectors. Unfortunately, the U-boat loss rate stayed high. The reason was simple; the British didn't have such a device after all.

So here was a case where deception, specifically dissimulative camouflage, failed because the enemy hadn't the sensors for their countermeasure to work against—there simply was no "channel" open between deceiver and the target.

In fact, the Germans had been hoaxed. Dr. R. V. Jones and his colleagues with RAF deception had planted the fiction that British scientists had developed an infrared detector to persuade the German Navy that this was the cause of their

---

56 Whaley, *Meinertzhagen's Haversack Exposed* (FDDC: 16 May 2007).
57 Hart-Davis (1974), 282-283; Holt (2004), 408-410, 819. Oddly overlooked by Howard (1990).

losses. Thus the real sensor (an improved radar) went undetected for several months.[58]

CASE 2.2.5:
Battle of the Sangro, Italy 1943

After landing on the Italian mainland on 3 September 1943, the Allies advanced slowly up the difficult spiny terrain. The Allied order of battle had General Alexander in overall command—of 15th Army Group—with the predominantly American Fifth Army under Lieutenant General Clark moving up the western half of the peninsula while the British Commonwealth Eighth Army under General Montgomery pressed up the eastern side.

On November 8th Montgomery reached the Sangro River. Alexander (and Churchill) now believed the time ripe for beginning his favorite strategy—a coordinated sequence of punches delivered from alternate sides of his line until a breakthrough was made by one or the other. This time Montgomery was to open the general offensive by striking across the Sangro. Originally scheduled to go on November 20th, Monty's attack was postponed twice (by rain-swollen rivers) to the 28th and then, as the weather suddenly cleared, went in one day early, the 27th. His main attack started the next day.

The delays and several abortive and partial assaults had evidently warned Field-Marshal Kesselring of the imminent offensive, for he used the time to reinforce that part of the line. Thus, on the 20th, Monty's assault force (V Corps) of three divisions had faced only a single inexperienced infantry division; but by the 28th the Germans had brought in advance elements of three experienced mobile divisions. Moreover, Monty's Intelligence was unaware of this major change in his enemy's forward deployment. Robbed of much if not all of the element of surprise, the battle raged for five uninspiring days. V Corps achieved only a 3 to 5 mile advance across 20 miles of front before being halted by exhaustion of its men and materiél.

The Battle of the Sangro is of special interest because Montgomery failed to gain surprise despite the use of comprehensive deception. Indeed, this battle is not only one of the rare cases where stratagem failed to yield surprise but may be the extreme instance where *elaborate* and *comprehensive* deception failed. The deception plan is therefore worth describing and analyzing in same detail.

---

58   R. V. Jones, *Reflections on Intelligence* (London: Heinemann, 1989), 141-143; R. V. Jones, *The Wizard War: British Scientific Intelligence 1939-1945* (New York: Coward, McCann & Geoghegan, 1978), 320-322; R. V. Jones, "The Theory of Practical Joking—Its Relevance to Physics," *Bulletin of the Institute of Physics*, Vol.8 (June 1957), 196; F. M. Hinsley, *British Intelligence in the Second World War*, Vol.3, Part 1 (London: Her Majesty's Stationery Office, 1984), 515-516.

First, although no overall deception plan embracing both Montgomery's eastern sector and Clark's western sector has been reported, it would have been quite out of character for Alexander not to have had some such notion in mind, particularly as he had planned Montgomery's punch from the right to prepare the way for Clark's left hook. Indeed, Sangro D-day was originally deliberately scheduled to follow by five days the end of Clark's first effort to break through—a threatening move that may have been intended to draw off German reserves earmarked for the east coast.

It is likely that if Alexander did have a macro-level deception plan it would have involved an exaggerated threat to the west coast. In any case, such a plan had at least one severe limitation. It was simply neither possible nor plausible for Alexander to use deception to mask shifts of large units between the two zones, the terrain and lack of adequate lateral communications precluding this. Consequently the Germans, having once determined the force distribution between Fifth and Eighth Army, would know they need not expect any sudden exchange between them. Moreover, as the Germans had better lateral communications networks they could easily keep pace with any known inter-army shifts. Finally, Alexander's superiority over Kesselring in local ground forces (18 divisions to 15) was not enough to force the German to abandon his low-risk defensive strategy, willingly trading one small river valley at a time rather than either fight a stand-still battle of attrition or risk entrapment by a too hasty commitment of units.

The only remaining strategic ruse available to Alexander—indeed the only one of any practical value under the circumstances—was to maintain a threat of further amphibious landings behind the German lines.[59] In this particular case Alexander may well have specifically decided against an amphibious feint on the western coast because he had, in fact, already planned a very real one-division landing at Anzio around December 15th.

If there was little that Alexander's 15th Army Group deceivers could do by way of effective strategic deception, Montgomery's Eighth Army staff planners were also rather limited in plausible tactical frauds with which to bedazzle the immediately opposing LXXVI Panzer Corps. Although there were four passable roads leading north, only one—the main coastal highway—was large enough and adequately surfaced to carry the traffic of an attacking corps. Thus Montgomery had, in effect, no choice but to take the obvious line of operations.[60]

To achieve this narrowly constrained and unpromising objective, Eighth Army employed its now stock set of hitherto (and subsequently) successful ruses in

---

59 Major-General W.G.F. Jackson, *The Battle for Italy* (New York: Harper & Row, 1967), 144-145.
60 Jackson (1967), 146-147.

a vain effort to misdirect the German defense to this inherently implausible alternative.[61]

Despite these efforts at tactical deception, the attack that went in on November 28th failed to gain surprise. Held to a slow and costly advance, Montgomery recommended that operations be suspended for the winter. Unfortunately I have yet to see any German documents that explain their assessment and reaction to the Allied deceptions.

CASE 2.2.6:
Plan COCKADE, The English Channel 1943

> Perhaps the most serious Allied deception failure[s] in Europe were operations "Starkey" and "Cockade," two hopelessly bungled efforts in 1943 to make the Germans believe that there would be an Allied invasion of Europe that year, in the hope of diverting German forces from the Eastern Front.
>
> — Gerhard L. Weinberg, *A World At Arms* (New Edition, Cambridge: Cambridge University Press, 2005), 557

COCKADE was a miserable and potentially dangerous failure for reasons spelled out in detail by Cruickshank, Mure, Brown, and Holt.[62] Consequently I will only outline the causes and the lessons learned.

COCKADE was the main Allied deception operation Europe in 1943. Believing themselves unable to open a Western Front that year to take pressure off the Russian Front, the British & Americans hoped to convince the Germans that their forces in England were planning to open a Western Front that year by an amphibious invasion of northern France. COCKADE had three components: STARKEY, a major feint across the English Channel; TINDALL, a purely fictitious ("notional") treat against Norway to hold German troops there; and WADHAM, another fictitious threat of American landings in Brittany.

Carried out that September, COCKADE and all its components failed completely in their purpose. The Germans noticed parts of these activities, remained unconcerned, and therefore took none of the actions that the Allied deception planners had intended.

---

61 These deceptions are described by Jackson (1967), 146-147; and Field-Marshal the Viscount Montgomery of Alamein, *El Alamein to the River Sangro* (New York: Dutton, 1949), 173.
62 Brown (1975), 317-327; Cruickshank (1979), 61-84; Mure (1980), 219-226; Holt (2002), index; Maxim (1982), 25-34; and *FM 90-2* (1988), Chapter 7, copying Maxim.

COCKADE had employed the full range of detection techniques—a mix of real & notional forces, false information from double agents, psychological operations, raids, and feints. However its execution had been too amateurish, too ill-coordinated, and too weak to be convincing.[63] Fortunately, the British psyops campaign—which involved one of the rare uses of an unwitting BBC—was terminated within 3 days when it was recognized that if French Resistance groups believed the deception they might take premature action.[64]

Frustrating? Yes. But a waste of time and effort? No. Viewed as a training exercise, it had led to the right "lessons learned"—particularly in better ways to handle double-agents. And Thaddeus Holt is probably correct in stating that the D-Day invasion "could not have run as smoothly as it did" without this rehearsal.[65] But that's small consolation for failure. I'm sure Holt would agree that while learning from failures is good, learning *before* failures is better.

CASE 2.2.7:
The Reluctant Pilot, England 1944

Major Ronald Wingate, the Executive Officer of LCS, recalled the following deception effort.[66] The scheme had been cooked up by, I presume, LCS and implemented by the RAF deception staff in early 1944. Conceived as a variation on both the old Haversack Ruse and The Man Who Never Was Ploy (Case 2.2.3), it was designed to feed Luftwaffe Intelligence a set of faked maps. For the messenger, the RAF selected a German POW who was a skilled fighter pilot and fanatical Nazi. The bait was the latest model Spitfire photo-reconnaissance airplane—a sleek, hot craft admired by pilots and desired by Luftwaffe Intelligence. The doctored maps were in the cockpit.

The recently captured pilot was brought to the RAF base at Leuchars in Scotland "for interrogation". At one point during the questioning, which took place at a desk by a window overlooking the airfield, the RAF interrogator excused himself to "see a man about a dog". Moments later an RAF ground-crew wheeled up the Spitfire outside the window, fueled the plane, started it up, and with shouts of "There's the NAAFI wagon!" rushed off for tea. There sat the sky-blue Spit, cockpit hatch open, its powerful Rolls-Royce engine revved-up and ready to go. And there sat the prisoner, alone in an unlocked room and not a soul in sight—patiently awaiting the return of his interrogator.

A clear case of a deception plan that literally as well as figuratively failed to get off the ground. The simple lesson learned is that the deception planners

---

63 Detailed assessments of this failure are Holt (2004), 492-493; Cruickshank (1979), 73-74, 80, 84.
64 Balfour (1979), 358.
65 Holt (2004), 493.
66 Wingate interview in Brown (1975), 522-523.

must plant their deception in a channel that is open to the target. Also we are entitled to speculate what might of happened if the RAF had tried again, this time with another enemy pilot with an escaper's mindset.

■ ■ ■ ■ ■

As mentioned earlier, the final and most serious cause of deception failure occurs if the enemy detects the deception itself. Although this has very rarely happened, it is a crucial category. Let us now look at the few known cases:

CASE 2.2.8:
Operation MI, Midway Island 1942

Japanese Admiral Isoruku Yamamoto made a confident effort to cap Pearl Harbor by a surprise *coup de grace* (Operation MI) against the bloodied and greatly outnumbered

U.S. Pacific Fleet at Midway on 3-4 June 1942. His cover plan had two prongs. Both were pre-D-day diversions, one in the North Pacific, the other in the South. The first was a raid on the Aleutians as part demonstration to mislead U.S. Admiral Chester W. Nimitz and part interdiction of any American bombers from there. The second comprised two diversionary attacks by Japanese midget submarines against Madagascar and Sydney, the latter to imply an imminent invasion of Australia. His goal was a surprise capture of Midway that would force the Americans to react. It was a bold Pacific-wide strategy designed to entice, trap, and destroy the few remaining American carriers.

Instead, Yamamoto himself met ambush and a stunning defeat by Nimitz whose cryptanalysts literally stole victory from Japan. By their solution of Yamamoto's detailed and comprehensive operations order of May 20th, Nimitz knew with complete confidence the Japanese fleet strength, deployment, strategy, timetable, and place of attack. Moreover, he even knew of the Japanese Naval General Staff's Aleutian diversion.[67] Although the American cryptanalysts had gotten no advance indications of the submarine diversions, the news on D-minus-3 of the actual sub attacks at Madagascar and Sydney were correctly interpreted as inconsequential sideshows.[68]

Nimitz' surprise counter-trap had bought 4 Japanese carriers, 322 aircraft, and 1 cruiser at the cost of 1 American carrier, 147 aircraft, and 1 destroyer. Yamamoto admitted that "In battle as in chess [shogi], it is the fool who lets

---

67 Rear Admiral Edwin T. Layton, *"And I Was There": Pearl Harbor and Midway—Breaking the Secrets* (New York: Morrow, 1985), 406-448; David Kahn, *The Codebreakers* (New York: Macmillan, 1967), 561-573, 603-604, 606; Walter Lord, *Incredible Victory* (New York: Harper & Row, 1967), 7, 9, 15-28, 76.
68 Layton (1985), 436.

himself be led into a reckless move through desperation." Nimitz summed up his victory in the statement:[69]

> Midway was essentially a victory of intelligence. In attempting surprise, the Japanese were themselves surprised.

That assessment was generous thanks to his intelligencers but overlooks the fact that intelligence is only a function, a means to an end. Even the best intelligence is worthless unless it is exploited, as Nimitz did in this case, by plans and actions.

CASE 2.2.9:
The Second German Counteroffensive at Anzio, Italy 1944
Another rare case of the deception plan detected.

After his first counteroffensive against the tenuous Allied beachhead at Anzio failed on 16-19 February 1944, the German C-in-C Italy, Field Marshal Albert Kesselring, prepared a second. Writing eight years later, he recalled that:[70]

> The lessons of the first attack were considered and measures of camouflage and diversion perfected, although I was not convinced this was necessary in so narrow an area.

This is a revealing statement. Kesselring was an air force officer with little grasp of the tactics employed on "narrow" ground and virtually none about the appropriate tactical deceptions. Fortunately, the plan for this battle was that of the field commander of the German forces (14th Army) ringing the beachhead, General Eberhard von Mackensen, who did appreciate the value of deception.

Von Mackensen's imaginative plan was accepted by Kesselring. It was to deploy his force in two spearheads, one on each beachhead flank, meanwhile seeking to deceive the beachhead commander, U.S. Army Major General Lucian K Truscott Jr and his VI Corps, that the attack would come straight through the center. To simulate the center buildup, Von Mackensen emplaced some 60 wooden guns and 180 dummy tanks in that area. This equipment had been specially made for the occasion in late February in the carpentry shops of the leading Roman motion picture studies at Cinecitta.[71] The illusion of strength given by this conspicuously displayed dummy materiél was enhanced by having the local units simulate attack preparations by extensive raiding and ill-concealed vehicular movement.[72]

---

69 Nimitz in E. B. Potter & Chester W. Nimitz (*editors*), *The Great Sea War* (Englewood Cliffs, NJ: Prentice-Hall,1960), 245.
70 Albert Kesselring, *A Soldier's Story* (New York: Morrow, 1954), 236.
71 Peter Tompkins, *A Spy in Rome* (New York: Simon and Schuster, 1962), 148-149.
72 Martin Blumenson, *Anzio* (Philadelphia: Lippincott, 1963), 147.

This was a good cover plan and well executed. It might have succeeded but proved worthless when Mackensen's counterattack began on February 29th. U.S. Fifth Army headquarters had already received reports of the complete details of the counteroffensive, including the ruses and camouflage. Moreover warning had come in time for Truscott (with appropriate air support from Fifth Army) to make full use of this intelligence to break the attack before it could develop thrust. This superb intelligence had been collected and sent along to Fifth Army by the lone OSS agent in Rome, Major Peter Tompkins. He had acquired it through the Italian Nenni Socialist underground intelligence service from a German traitor inside Kesselring's own headquarters.[73]

Denied surprise and meeting fierce resistance, Kesselring wisely cancelled his broken attack the next day. There was a lesson to be learned, but the German Field-Marshal didn't learn it. Unaware of the security breeches at his headquarters and at Cinecitta, defeat only confirmed his contempt for tactical deception.[74]

CASE 2.2.10:
Operation ELEPHANT, Normandy 1944
The 23rd Hq Special Troops has its first and only total failure.

Following two aborted deception plans,[75] the American 23rd Headquarters Special Troops mounted its first deception operation on 1 July 1944. The mission of Operation ELEPHANT was to simulate the 2nd Armored Division remaining in reserve while the real 2nd Armored moved into battle.[76]

ELEPHANT employed 37% of the unit's personnel (395 officers and men). As each element of the 2nd Armored moved out from the reserve area, the 23rd tried to replace the departed vehicles with inflatable rubber dummies, one-for-one in the case of tanks. In some cases this effort was successful; but in others the replacement was late or incomplete due to poor coordination—only a single deceptive radio was used and no sonic deception was employed because the sonic deception company was still in England.

No effort was made to simulate the departed troops with counterfeit shoulder patches, bumper markings, or command post signs. It was an altogether half-ass effort.

---

73 Tompkins (1962), 148-149; Fred Sheehan, *Anzio* (Norman: University of Oklahoma Press, 1964), 158-159.
74 See Whaley, *Stratagem* (1969/2007), Example B33 ("Second Counteroffensive at Anzio"), pp.365-366 in 2007 reprint. Labeled a "failure" by Maxim (1982) and, copying Maxim, *FM 90-2* (1988), Chapter 7.
75 The unnamed plan of June 6th and TROUTFLY of June 7th.
76 *Official History 23rd Headquarters Special Troops* (1945), 7-9.

Meanwhile the real 2nd Armored Division had reached the front, four miles away, moving without camouflage. Its artillery was firing and some of its infantry was already in the line where it was assumed to have been properly identified by the Germans. The operation was cancelled on its 4th day when the fresh 3rd Armored began to enter the dummy reserve area.

The unit's official historian identified the following "lessons learned", which were the only positive results of this otherwise failed operation:

- Need for close coordination between the deceivers and higher headquarters.
- Need for close coordination with the units being simulated.
- Need for what later was called "Special Effects, the simulation of the real unit's "personnel by such spoof devices as counterfeit shoulder patches, phony major generals, empty gas convoys, and the spreading of rumors. The target of "Special Effects" was the behind-the-lines enemy agents and loose-tongued civilians.
- Recognition that the 23rd's capability was pretty much limited to simulation of no more than a single division unless additional troops were attached.[77]
- The value of dummy equipment was negligible unless the enemy were flying aerial reconnaissance.

CASE 2.2.11:
Twenty Committee Fails at "Coat-Trailing", Germany 1940-43

An interesting type of failure can and often does occur in every effort to develop deception assets. As this relates only indirectly to failed deception operations, two examples should be sufficient to make the point. The first illustrates the frustrations of developing double agents; the next (Case 2.2.12) the problems of introducing specific devices for deception. But first the double-agent problem:

The British Twenty ("Double-Cross") Committee had proved extraordinarily adept in WW2 at taking captured German agents, turning them around, and working them back as double agents against German Military Intelligence, the Abwehr. In addition, the Double-Cross Committee made several attempts to recruit and plant new agents on the Abwehr. Nearly all these latter efforts failed. The Committee's Chairman analyzed this problem:[78]

---

77 Whaley, "The one percent solution: costs and benefits of military deception," in John Arquilla & Douglas Borer (editors), *Information Strategy and Warfare* (New York: Routledge, 2007), 147-149.

78 J. C. Masterman, *The Double-Cross System in the War of 1939 to 1945* (New Haven: Yale University Press, 1972), 17-18.

Double agents ... should *not* be created. If a person offered himself and could satisfy us that he had in fact been approached and recruited by the Germans, he could probably be used with advantage. On the other hand when we tried to throw an agent in the way of the Germans with a view to persuading them to recruit him, we almost always failed. This rather curious fact was in some ways the most important practical lesson which we learned in the theory of double-cross work. Cases occurred in which an individual appeared to have been specially designed by providence to become a double agent. Imagine an Englishman who has spent much of his early life in Germany, and who has business interests and relatives in that country; suppose too that he has real reason to dislike the British government—he may, for example, have been cashiered from the army or sentenced in the courts for some illegal transaction, or he may have become attached to some fascist organization in time of peace; in spite of all this he has ostensibly remained a loyal British subject. Such a man, it might be supposed, has only to be encouraged to appear in a neutral country for the Germans to attempt to recruit him. In point of fact, such bait was rarely taken. "Coat-trailing" was usually a failure.

The main reason became clear later. It was an Abwehr tradition for each of its officers to recruit his own agents. British M.I.6, who were responsible for the planting, obviously failed to do their coat-trailing either in the right places or with the right Abwehr officers.

One specific example of coat-trailing was Plan MICAWBER, designed in 1941 to induce the Abwehr to accept a British agent planted in Censorship. It failed when the Germans declined the bait. While this particular operation failed in its purpose of deception, it had the solid fringe benefit of collecting valuable intelligence. As the Chairman of Twenty Committee observed: "... it should not be forgotten that the essence of counterespionage is prevention, and the failure of this plan helped to satify us that the Germans were not in fact making any real attempt to use censorship leakages for their own purpose." In other words, this effort to open a specific deception channel to the enemy failed because the Germans were simply not interested in that particular source of intelligence, as they had been in the previous world war.[79]

Although all this was a bit frustrating for Twenty Committee, it had its "bright side", as its Chairman wryly noted: "Though they would not take a first-rate

---

79   Masterman (1972), 86-87.

article from us, the Germans showed themselves more than willing to push a second-rate article of their own.[80]

CASE 2.2.12:
Col. Wintle's Device, Cyprus 1942

The British island of Cyprus in the Eastern Mediterranean had a serious invasion scare in late July 1942. At that time the "A" Force deception planners had convinced German Intelligence that Cyprus was the staging point for a notional effort to recapture Crete and the fear of German preemption was real.

As an Intelligence officer with British Ninth Army involved recalled:[81]

> It was during this scare that the great Colonel Wintle arrived from Cairo with the "Road Deceptive Wintle", a cardboard extension which, spread across the ravines on the precipitous mountain roads of Cyprus, was designed to lure German soldiers to their destruction. This was an example of an 'A' Force proposition which was 'completely still-born', and for which no final use was ever found 'so that the effort, ingenuity and hope put into it was wasted'.

As examples of art imitating life, I am reminded of a later Charles Adams' *New Yorker* cartoon and an even later Roadrunner-Wily Coyote cartoon film, both of which employed variants of this road trap.

Such examples of technological invention—some realistically clever, others mere comic fantasies—are common among military deceptionists. Unlike Wintle's, most of the latter never get off the drawing board, fortunately.

CASE 2.2.13:
MacArthur's Invasion of Luzon, Philippines 1945

General Douglas MacArthur was rather adept at deception, as proven repeatedly throughout his career. He (and his staff) proudly claim that their deception plan to cover the invasion of Luzon succeeded. It didn't.

The invasion of the main Philippine island of Luzon (Operation MUSKETEER III) on 9 January 1945 was the largest single amphibious assault of the Pacific war to that time. The Americans would put some 203,000 troops put ashore in 10 divisions and 5 regimental combat teams to join the equivalent of 1 division and 2 regiments of local Filipino guerrillas.

---

80   Masterman (1972), 17-18.
81   Mure (1980), 109. Further details of Wintle and his strange invention are in Mure (1977), 39-40.

The Japanese defenders—the 14 Area Army—mustered 275,000 troops (123,000 more than MacArthur's not always perfect G-2 had overconfidently estimated). Moreover, this formidable force was commanded by one of the more imaginative Japanese commanders, General "Tiger" Yamashita who had been given that critical command in September in correct anticipation of the nearness of MacArthur's return to the Philippines.

A coordinated, inter-theater deception operation was tried for the second time in the Pacific by the Americans as a preliminary to MUSKETEER III. Conceived by MacArthur and planned and coordinated by the JCS, it was, as the official historian summarized:[82]

> ... a Pacific-wide deception program to make the Japanese believe that the Formosa-Amoy area, rather than Luzon, would be the next major Allied target after Leyte.

Later, MacArthur's own command launched a three-part deception plan to convince Yamashita that eventual landings on Luzon would take place on its southern beaches, to draw the Japanese defenders away from the real target site far to the northeast in Lingayen Gulf.[83] MacArthur and his admiring staff—Willoughby, Whitney, and Mashbir—all claim these operations were entirely successful in focusing Yamashita's attention on southern Luzon. For example, MacArthur claimed that.[84]

> The Japanese, as I had hoped, were deceived, and moved troops south in anticipation of an attack there. Not until we had actually landed at Lingayen did General Yamashita move his center of gravity to the north. He established his new headquarters at Baquio. The deception [by] the [American] Eighth Army had been remarkably successful.

Similarly, Major General Courtney Whitney asserted:[85]

> These tactics of deception worked. The enemy was tricked into rushing a division of troops from its northern force to Batangas and another division to Bataan.

On the contrary. These statements illustrate both the pitfall of mistaking coincidence for cause-and-effect as well as the too common failure of intelligence services to test their initial estimates against subsequent evidence. In fact, Yamashita had very shrewdly fathomed MacArthur's intentions and

---

82 Robert Ross Smith, *Triumph in the Philippines* (Washington, DC: Office of the Chief of Military History, 1963), 53.
83 Details and sources are in Whaley, *Stratagem* (1969/2007), Case A52 (Luzon). See also D. Clayton James, *The Years of MacArthur*, Vol.2 (Boston: Houghton Mifflin, 1975), 616.
84 General of the Army Douglas MacArthur, *Reminiscenses* (New York McGraw-Hill, 1964), 239.
85 Courtney Whitney, *MacArthur* (New York: Knopf, 1956), 180. Compare Smith (1963), 241, 311-312.

correctly predicted the main landing would come in Lingayen Gulf. He had originally placed 36,000 of his troops there to trap the invaders. Now he adopted an unprecedented and therefore unexpected strategy. He knew he could expect no serious support from the recently sacrificed Japanese fleet and airforce. He realized that the enemy's capture on December 15th of nearby Mindoro Island with its four excellent airfields would give the Americans overwhelming force at their beachhead. Therefore, he chose to give them their landing and use his time and resources to prepare a redoubt in the northern mountains, conducting a static defense there with only minor delaying actions elsewhere on the island.

Accordingly, Yamashita redeployed his Lingayen Gulf troops to the north and moved his own headquarters north from Manila to Baguio. And his alleged "reinforcement" of Batangas—if it took place at all—was, I presume, only a result of his pre-invasion decision to evacuate the Manila garrison south for a delaying operation against a U.S. drive down from Lingayen. MacArthur's G-2, Willoughby, predisposed to believe the Japanese would imitate MacArthur's historic withdrawal into the bastion of Bataan peninsula, estimated 13,000 Japanese troops at Bataan when, in fact, only 4,000 were in the whole region. Moreover, their orders were to retreat to the north-west mainland rather than withdraw into the barren Bataan peninsula. Indeed the alleged "division" reinforcement in late December comprised only a partial regiment and was merely the fortuitous result of the cancellation of that regiment's orders to reinforce Leyte.[86]

The initial landing on Luzon on January 9th took place on the northwest coast across the southern beach of Lingayen Gulf. The landing was unopposed. Substantial *tactical* surprise of time and place had, however, been obtained because Yamashita's planners had categorically ruled out that specific beach in the bay as too inferior a landing site and because Yamashita himself believed the invasion was at least a fortnight away.

CASE 2.2.14:
Operation DECOY, Korea 1952

This case is only briefly summarized here, as it's already received detailed treatment and documentation elsewhere.[87]

Being a peninsula, Korea is a particularly attractive target for surprise amphibious outflanking attacks. General MacArthur had grasped this simple but then doctrinally unpopular concept within a fortnight of the outbreak of

---

86 Whaley, *Stratagem* (1969/2007), Case A52 (Luzon); William Manchester, *American Caesar* (Boston: Little, Brown, 1979), 406; D. Clayton James, *The Years of MacArthur*, Vol.2 (Boston: Houghton Mifflin, 1975), 625-626.
87 Whaley, *Stratagem* (1969/2007), Case A61 (The Kojo Feint).

the Korean War in 1950 and proved his point three months later at Inchon. Although Inchon would be the only amphibious end-run of the war, it had made the North Korean & Chinese commanders sufficiently nervous for them to henceforward keep substantial reserve forces tied down to cover any future landings. Nevertheless, one major American feint, Operation DECOY, proved a general failure.

Although this feint against the major west coast harbor of Kojo was fairly elaborate, it failed to draw any noticeable reaction. While several explanations have been offered, I suspect the most likely one to be simply that the 13 months since Inchon had been more than enough time for the Korean & Chinese to satisfy themselves that such contingencies were sufficiently covered.

2.3. Backfire

> A feint which did not fully fulfill its purpose would have been worse than useless.
>
> — General Monro, despatch at Gallipoli, 6 Mar 1916

It is surprising rare for any military deception operation to fail badly enough to actually harm the perpetrator. In fact I have been able to find only four *battlefield* deceptions that did so—Case 2.1.2, described above under the Partial Failure section, and three others in 10 cases that follow in this section.

This section also includes examples of camouflage—both dissimulative & simulative—that backfired, harming their perpetrators. Other examples can be found, particularly in the areas of military and political espionage, where such failures are not uncommon.

The cause of all cases of backfire is simply bad tradecraft and, consequently, avoidable through sound theory & proper training.

CASE 2.3.1:
The German Summer Offensive, Russia 1942

By February 1942 Hitler was deeply involved in planning his forthcoming summer offensive in Russia. As before, he alone planned grand strategy. However, since December when he had replaced Field-Marshal Von Brauchitsch as Commander-in-Chief of the Army, Hitler had also extended his secretive and capricious style to the central operational planning. Thus the summer offensive plans underwent several sudden and bewildering changes in both objectives and timing that confused his own German military staffs more than the enemy.

This proliferation of mutually exclusive plans made it impossible for the Supreme Command of the Armed Forces (OKW) to develop effective supporting deception plans. For example, sometime in mid-February, OKW (and the Army Staff) issued written instructions with the concealed purpose of misleading the Soviets. One outcome was that the Navy ordered preparations for operations in the Black Sea, thinking it to be a feint.[88] However, these efforts soon backfired as Hitler gradually shifted his real target south, turning Moscow and Leningrad into increasingly secondary targets.

Hitler took until April 5th to make up his mind. That day he issued War Directive No. 41, formally specifying that the offensive would strike through the southern front. To mask this intention the OKW launched a major deception campaign to convince the Soviets that the offensive would be aimed straight at Moscow.[89]

When the offensive opened on June 28th the Soviets were reasonably forewarned.

Their Intelligence had been receiving enough good reports on German strategic plans and order of battle to be reasonably set. Moreover, on D-minus-9 they had captured the complete enemy battle plan for the opening drive. Consequently they were able to avoid a disastrous battle by—for once—a somewhat orderly withdrawal behind the Don. The fact that the original German deception operations had pointed to the south may have even helped the Russians concentrate there.[90]

CASE 2.3.2:
Operation BREST, France 1944

A curious blunder in planning was made by General Bradley's 23rd HQ Special Troops preceding the final successful assault on the German-held fortress of Brest. Operation BREST ran from 20 to 27 Aug 1944, employed a good mix of radio, sonic, and visual ploys to convince the German garrison of the direction of the forthcoming American tank offensive. Unfortunately, the 23rd's planners had neglected to coordinate with the commanders of the real combat units on the ground. Consequently, their deception succeeded too well—it alerted the Germans to the *real* point of the armor attack. Thus Operation BREST,

---

88 Walter Warlimont, *Inside Hitler's Headquarters, 1939-1945* (New York: Praeger, 1964), 229-230, 615n6.
89 For details on German deception see Whaley, *Stratagem* (1969/2007), Case A34 (Summer Offensive in Russia).
90 For Soviet Intelligence assessments see Whaley, *Stratagem* (1969/2007), Case A34 (Summer Offensive in Russia).

which Holt had mislabeled as merely inconclusive, was actually merely incompetent—one where the American's cover story indicated the real place of their attack, which the German's had already presumed.[91]

CASE 2.3.3:
The Indaw Raid, Burma 1944

On 24 March 1944 Brigadier Bernard Fergusson led one column of the 1,800 troops of his Chindit 16th Infantry Brigade in a long-range attack intended to capture the Japanese airfields at Indaw. The main part of the deception plan was to put out the word in the Burmese villages through which they passed that their target was the railway town of Pinwe, some eight miles away from Indaw. (The other part was to position his reserve column during the final approach march on the opposite side of Indaw than the real attack. This second part aborted, as we shall see, when the whole operation was cancelled.)

Six hours into his march at the village of Nyaunggon, Fergusson received an astonishing report from one of the villagers. The man said that "nine hundred British troops passed through here two days ago on their way to attack Indaw." Fergusson was puzzled:[92]

> Who could these be? The Leicesters [one of the Brigade's battalions], I knew, had not come this way; and even if they had they would not have announced their intention like this. ... With a sinking heart I recognized the description of the troops as of two Chindit columns, and the description of the interrogation as the old technique of selling a dummy to the Japs by asking the locals a string of questions about a false objective.
>
> It was easy to see what had happened. Two columns of some other Brigade, heading probably for Pinlebu, had crossed my front, using an imaginary attack on Indaw as their cover plan. They had presumably not been briefed that my Brigade was, in fact, going for Indaw, and that what was an admirable cover plan for them was the worst possible from the point of view of my own task. I confirmed, long afterwards, that this was precisely what had happened. It was another example of the old business of excessive security defeating its own object; the concealment of the main plan from subordinate commanders resulting in its chances of success being compromised.

---

91   *Official History of the 23rd Headquarters Special Troops* (1945), 12-13.
92   Bernard Fergusson, *The Wild Green Earth* (London: Collins, 1952), 103.

Fergusson could not postpone his attack and could only hope that the Japanese "would probably be getting conflicting reports, and perhaps this new one would only muddle them further." In fact, it was the British brigades that had muddled the picture for each other. Fergusson's columns were not the surprisers but the surprised; and, after several fruitless but costly engagements, he wisely decided to break off his attack and withdraw.

CASE 2.3.4:
Dissimulative Camouflage Fails, Japan 1944

> Camouflage design is worth tons of camouflage correction. Prevention is better than cure.
>
> — Solomon J. Solomon, diary, 20 Aug 1918[93]

The Japanese were unexcelled at individual and jungle camouflage but astonishingly failed to understand how to apply these principles to large structures. Instead of using color and pattern to tone down these structures and blend them into their surroundings, the Japanese camoufleurs went the opposite direction, dazzle-painting many key installations in highly visible designs unrelated to their surroundings. Oil tanks near Kure were painted with bold rectangles and diagonal slashes; hangers at the Yokosuka seaplane base and factories in Kobe and Osaka were painted with large, vivid, geometric patterns; some railway trains were covered with brightly colored blobs and patterns.

When in November 1944 General Curtis LeMay's 20th Air Force began regular mass bombings of targets in Japan, his B-29 navigators and bombardiers found this odd Japanese camouflage actually made their targets more visible. As one navigator observed, "they made our work easier for us." Only in the final months of the war did the Japanese camoufleurs begin to correct their error in camouflage theory.[94]

This is an example of failure to create plausibly dissimulative deception.

---

93 As quoted in Olga Somech Phillips, *Solomon J. Solomon: A Memoir of Peace and War* (London: Herbert Joseph, 1933), 203.
94 Seymour Reit, *Masquerade: The Amazing Camouflage Deceptions of World War II* (New York Hawthorne Books, 1978), 190.

CASE 2.3.5:
The Tell-Tale Photographs, Germany 1944- 45

All Allied airmen shot down and captured in Nazi Occupied Europe in WW II were interrogated by Luftwaffe Intelligence. Some had managed to evade capture long enough to disguise themselves in civvies. If these men failed to answer questions beyond the old name, rank, and serial number bit, the interrogators told them that they would be shot as spies unless they gave enough details to prove they were airmen. These ex-evaders did not know that their interrogators knew before seeing them that they were not spies but airmen. Moreover, the interrogators already knew whether they were RAF or USAAF, whether they flew bombers or fighters, and even from which airfield they had flown their final mission.

The interrogators got this information from Sergeant Bert Nagel, the Luftwaffe's expert analyst of personal possessions taken from downed Allied airmen, dead or alive. Nagel had gleaned these particular details from the "escape photos" all American and British aircrew were required to carry on missions. They were intended for use by the French, Belgian, and Dutch undergrounds in preparing false identity papers for the escapee. These small ID photos were taken at the airman's home base, showing him in civilian clothes, and issued in a set of five photos wrapped in a cellophane-wrapped packet. This was a small telltale itself when, as in some cases, the prisoner had been caught with the packet intact.

The complete giveaway, however, was the photos themselves. Each airbase photographer had used a stock set of second-hand clothing (jacket, shirt, and necktie), designed his own background, and had his own distinctive way of cutting and trimming the final prints. Nagel "loved to work out his scheme systematically and soon he knew all the suits, shirts and ties, stripped or checkered," from each Allied airfield. In this way he could identify each POW's bomber or fighter unit and hence his base from just a quick look at the man's picture. This knowledge was, of course, also part of the advance information that the interrogators could leak to the ordinary POWs to give a simulated impression of omniscience.[95]

The lesson learned—but not in time for that war—was to avoid such "tells" in deception design. In this case they formed a systematic pattern that was easy for a clever enemy analyst to detect.

---

95 Raymond F. Toliver, *The Interrogator* (Fallbrook, CA: Aero, 1978), 56-57, 145, 338-339.

CASE 2.3.6:
The "KGB Poisoners" at Radio Free Europe, Munich 1956

Major Dr. Ladislav Bittman gives us a fascinating insider's account of the operations of the Czechoslovak Interior Ministry's Disinformation Department where he served in 1964-66 as Deputy Director. The next two examples are cases that backfired. The first was one Bittman knew of unofficially at the time and officially after the fact.

Located in Munich, RFE was then America's ostensibly private but actually CIA's principal propaganda agency directed at the Soviet Bloc countries in Eastern Europe. In 1956 the Western media headlined RFE's (i.e., CIA's) claim that the Soviet KGB had tried and failed in an attempt to poison the entire staff of Radio Free Europe while having lunch in their headquarter's cafeteria. The Communist propaganda machinery (i.e., the KGB) blustered that it was just another CIA plot to discredit the Soviet Union. Both the CIA and the KGB propagandists lied. The truth was more complicated and much more clever—on both sides.

The plan had originated with the Czechoslovak Interior Ministry (i.e., "secret police"). The idea was to put potent doses of laxative crystals in the RFE cafeteria saltshakers. The purpose was, as Major Bittman later explained, two-fold. First, a kind of scatological "practical joke" designed "mainly to amuse themselves." Second, "to create an atmosphere of fear among RFE employees". The assumption was that the victims, who were mostly East European refugees, would realize that "Big Brother" was watching their "treason" very closely, so closely that he could have killed had he so chosen. It was to be a covert operation with no direct evidence of enemy action, just enough suspicion to sew fear and panic inside RFE. The operation was approved by the Ministry and assigned to Jaroslav Nemec, an intelligence officer stationed in Salzburg under consular cover. Nemec had been picked because he already ran several of the Ministry's well-established agents working among RFE's service staff. Nemec selected, briefed, and sent one of these agents on this mission.

The scheme collapsed when Nemec's agent balked at the last minute and reported the plot to RFE officials. They let the saltshakers be loaded and put on the cafeteria tables,[96] stopped the game at that dramatic point, and then publicized the affair. Newspapers, radio, and television reported RFE's deliberately exaggerated version: stressing that atropine caused delirium, convulsions, coma, or death; named Nemec; and attributed the "poison" plot to the KGB. Thus it was the CIA/RFE that turned this little psychological warfare battle to its own advantage.

---

96 Most accounts say that only a single previously loaded saltshaker was brought to RFE by the Czech agent.

Only later did the Munich public prosecutor establish that the amount of atropine was too small to cause serious harm much less be life-threatening. But the truth came too late to undercut the CIA's propaganda coup.[97]

CASE 2.3.7:
An R&D Backfire, Czechoslovakia 1960s

During the mid-1960s the Czechoslovak Interior Ministry's Department D avidly sought ways to deceive Western intelligence services about the directions that Czechoslovakian and Soviet military and intelligence technology were moving. As we shall see later, most aborted.

Unable to gain the necessary cooperation of prestigious scientists, Department D finally made one effort to launch their first R&D deception, running it as a strictly in-house job. They drew only on the technical advice of their own electronics experts, These experts suggested a plausible but unfeasible new type of electronic device to jam nearby radio communications. The ostensible target would be the West German military radio command and control net; the real target was West German intelligence. Department D then had its carefully cultivated rumor networks in West German electronics circles spread the story that Czechoslovak scientists had secretly perfected this device.

Initially Department D was pleased to learn that West German Military Intelligence had "bought" this fiction and were hastily diverting some of their best scientific brains and much research funds to finding an effective countermeasure to this nonexistent jamming threat.

Then the whole clever scheme backfired. Three Czechoslovak government labs also started work on the same worthless project. Having learned from German colleagues at international conferences of the research in West Germany, they had decided on their own that Czechoslovakia would need a defense against this (imaginary) threat to her own military communications. It was too late for Department D to warn them off without causing a major bureaucratic scandal, so the labs were simply allowed to continue their efforts until satisfied that the original jamming device was impractical."[98]

97　Ladislav Bittman, *The Deception Game: Czechoslovak Intelligence in Soviet Political Warfare* (Syracuse, NY: Syracuse University Research Corporation, 1972), 11-12. Also my interviews in 1969 with Major Bittman and Dr. William E. Griffith who had been RFE Political Director at the time.
98　Author's interview with Maj. Dr. Ladislav Bittman, 1972.

CASE 2.3.8:
Covert Operations, USA 1987

A special category of failure is often seen in political and political-military deception ops, particularly those in support of covert operations. The unraveling of the Iran-Contra connection in 1986-87 is a familiar case in point.

Why do covert political-military operations have such a high failure rate? I suggest two reasons. First, they tend to be much more complicated in design than either purely military or purely political ones. The combination generates a correspondingly larger number of discrepancies for the target to detect. Second, they tend to extend over longer periods of time. That gives the target more time to find and put together the various clues.

In the Iran-Contra case, both of these counter-productive circumstances had been built in. It was Marine Lieutenant Colonel Oliver North's first covert assignment and, while he would later give the U.S. Congress a lucid briefing on the nature of such jobs, he had clearly lacked the experience to know when to terminate his own operation. His effort then had been to preserve secrecy. It should have been to anticipate a point before secrecy would be apt to collapse and end down the operation at that point. If one wants deniability to be truly plausible, the operation must be shut down before security fails.

Another prominent case where a covert operation was blown in mid-course occurred in the late 1960s when the American press finally unraveled the CIA's covert subsidy of various international leftist but anti-Communist publishing houses (including Praeger), labor unions, student groups, etc.[99]

CASE 2.3.9:
Jody Powell and the Iranian Rescue Mission, 1980

Jody Powell was Press Secretary throughout his old boss Jimmy Carter's tenure as President of the United States. In 1975 British journalist Anthony Cave Brown published his *Bodyguard of Lies*, the first detailed account of the British deception operations that culminated in the Allied D-Day landings and breakout in Normandy. It became a best-seller in the USA.

Jody Powell read a copy. It was a revelation for him and so, sometime after 1977 when he accompanied President Carter into the White House, Powell praised its advocacy of military deception to all who would listen.

His chance to follow his own advice came in April 1980 when the administration was about to launch a rescue mission of the 52 hostages being held in the U.S. Embassy in Tehran, Iran. At that time Press Secretary Powell was asked by

---

[99] A story best told by Hugh Wilford, *The Mighty Wurlitzer: How the CIA Played America* (Cambridge, MA: Harvard University Press, 2008); and most intelligently critiqued by Michael Warner in *Studies in Intelligence*, Vol.52, No.2 (2008), 71-73.

Jack Nelson of *The Los Angeles Times* if the government was planning such a mission. As Powell publicly admitted some 5 months later, "And I said, 'No.' And then I went on at some length to try to make convincing arguments about why we weren't about to do that." Powell added that had he replied merely "no comment," it would have tipped off Mr. Nelson. He admitted he was still disturbed about his words, but had decided, "I did not have the right to endanger people's lives." In conclusion, Mr. Powell said, "If you are prepared to lie to protect people's lives, not to be too flip about it, but it had better be a good lie."[100]

The ill-named Operation EAGLE CLAW launched on 24 April 1980. Later that day at DESERT ONE, a remote & secret staging ground deep inside Iran, the operation aborted. Through a series of accidents during a blinding sandstorm one of the Sealion helicopters smashed into a C-130 transport, destroying both aircraft and killing 8 servicemen. Five other helicopters were abandoned. It proved an immediate, public, and embarrassing failure. It contributed to Carter's failed bid that November for a second term in an election that swept Ronald Reagan into the White House.[101]

All hostage rescue missions are high risk affairs. But EAGLE CLAW was almost assured of failure by a plan comprised of four components, each of which was deeply flawed. Moreover, because all these components were interdependent, the final likelihood of failure was governed by the ruthlessly inflexible Calculus of Probabilities Rule. The diverse component parts were: 1) A small CIA paramilitary team in Tehran; 2) a large Navy fleet carrier task force in the Persian Gulf; and in the Iranian desert, 3) Marine helicopters; 4) Air Force transports; and 5) Army Delta Force commandos.

Although each of these components posed only a moderate risk, their synergy was undermined by the following factors:[102]

- **Amateurish overall concept.** Here we see Press Secretary Powell's romantic derringdo influence at the Commander-in-Chief level. The only antidote at that level is expert advice. At that time in the White House, expert advice was weak to missing and consequently either overlooked or discounted.

---

100 Elisabeth Bumiller, "New Slogan in Washington: Watch What You Say," *New York Times*, 7 Oct 2001. Jody Powell, *The Other Side of the Story* (New York: Morrow, 1984), 252-314, which cover disinformation and deceit as manipulated by presidential press secretaries and gives Powell's soul-searching identification of the circumstances when a press secretary is justified in lying.

101 Paul B. Ryan, *The Iranian Rescue Mission: Why It Failed* (Annapolis: Naval Institute Press, 1985); Mark Bowden, *Guests of the Ayatollah: The First Battle in America's War with Militant Islam* (New York: Atlantic Monthly Press, 2006); *Wikipedia*, "Operation Eagle Claw" (accessed 2 Apr 2010).

102 This is my assessment. However, it largely agrees with the conclusions & recommendation of the official post-action inquiry, the so-called Holloway Commission.

- **Weak interagency coordination during the planning.** The most effective standard antidote is war gaming, as the Israeli General Staff proved in the 1976 Entibbe Airport Rescue Raid. But only, as with the Israelis, when a permanent gaming capability—ready to design & run a game on one day's notice—is available at senior staff level.

- **No *coordinated* training of the separate units.** Not even the cross-training of the unit commanders that could at least have been easily simulated in a war game.

- **No preplanned Out.** From the point when the operation was aborted, only the exfiltration of the Tehran team went according to plan. The catastrophe at Desert One had not been anticipated. While its specific cause—the sudden and unusually severe weather—was only a remote possibility, the several other more likely accidents were given too little weight. So, there was no plan to rescue the rescuers with the *predictable* result that when the two aircraft went up in flames it became that worst of all military retreats—a frantic *suave qui peut*, every man for himself. Whenever the consequences of total failure—however remote it may be—are enormous, it is inexcusable to have no preplanned Out.

CASE 2.3.10:
British Black Propaganda Operation MÖLDERS Backfires 1941

> The "black" artist must have for his intended victim that strange quality which we call "empathy" and the Germans call "Einfühlungsvermögen".
>
> — Sefton Delmer, *Black Boomerang* (1962), 297

As of 22 Nov 1941, the German Luftwaffe's top fighter ace with over 100 kills (15 in the Spanish Civil War, 53 on the Western Front, & 32 on the Russian Front) was Col. Werner Mölders. Although he was an avid Roman Catholic and openly non-Nazi, to preserve this national hero, he was reassigned to a safe senior administrative post. Ironically, he was killed that day while traveling as a passenger on a transport plane that made an accidental crash landing.

The British propaganda machine, the Political Warfare Executive (PWE), saw an opportunity to heat up the existing suspicion & friction between the German Roman Catholic church and the Nazi regime. Accordingly, PWE forged a letter ostensibly to a Catholic official in which "Mölders" complained bitterly of "the godless ones" and their "slime of lies, injustice and perversion." This letter, reproduced on counterfeit Luftwaffe stationary in thousands of copies was then scattered over Germany by the RAF. One implication was that they might have been dropped by pro-Mölders anti-Nazi German night-fighter pilots on their sorties against RAF bombers.

The German response was nationwide. The regime denounced the letter as a forgery. But church officials—Protestant as well as Catholic—judged it genuine. And the actual method of delivery was not unmasked, seemingly not even by the regime. So, the deception had been believed by the civilian target audience.

However, while believed, this clever deception operation failed in its goal. Intended to provoke greater dissension between German church and state, the actual result was a backfire. The Church, from the hierarchy down, became even more passive, more reluctant to confront the Nazi regime or even support opposition by others.[103]

2.4. Turnaround vs Turnabout: Blunder vs Entrapment

It is important to distinguish unintended self-induced backfire or blowback from entrapment—when your enemy deliberately draws you into a trap. The author's previous paper was titled *Turnabout: Crafting the Double-Cross* (FDDC: Jan 2010). There turnabout required that the intended victim perceive the deception and turn it back upon the deceiver—the hunted becoming the hunter. But here, *turnaround* requires only that the hunter fall into her own trap, that the sapper step on his own landmine, that the retreating soldiers trigger their own booby-traps, or comedians slip on their own discarded bananas.

The French have contributed an interesting concept to military deception theory, that of the *abcès de fixation*, an extreme sub-category of what Liddell Hart had earlier called the "luring defensive". The idea is to create a baited trap, a target so attractive that the enemy will assault it at a cost in casualties and materiél far beyond its real worth. It is not merely a decoy set up in some irrelevant place; it is a lure designed to become the central cauldron of the battle. To create it, the deceiver must make the enemy perceive a high payoff for himself by taking a target that, in fact, is unimportant to either combatant. Unlike the decoy, the lure is vigorously defended but on a killing ground of the defender's choice. Its purpose is to create a battle of attrition where the casualty ratios greatly favor the defender.

This is a clever idea that has succeeded on occasion and is an interesting option for commanders and their deception planners. But it can backfire. Here are

---

[103] Sefton Delmer, *Black Boomerang* (London: Secker & Warburg, 1962), 136, 138-139; Balfour (1979), 248; Deception Research Program (1982), 21-22.

two instructive examples where the deceptive lure was initially successful but eventually backfired—disastrously.

CASE 2.4.1:
Verdun, a Lure that Turned Around, Western Front 1916

The German Commander devises a cunning killing-ground that goes awry because he fails to inform his superiors.

At Verdun mankind achieved a new plateau of emotional horror and intellectual and moral bankruptcy. The German and French military and political leaderships collaborated to invent this mindless, fruitless, "passionate prodigality". Other battles (the Marne and the Somme) had run up bigger butcher bills. And still others (Passchendaele) had by logistical triumphs managed to develop more efficient killing ground. But only Verdun so perfectly simulated hell.

Yet even this most brutally stupid of battlefields produced a few imaginative flashes of surprise and deception. Verdun had actually begun in a moment of tactical surprise, the only event in that terrible battle that won any tangible trophy for the Germans. Also it was the only occasion after the Western Front got bogged down into a trench siege war where the Germans gained ground for proportionately fewer casualties. The fact that this episode occurred just outside the somber central drama that rightly dominates the minds and emotions of Verdun's memorists and historians explains why it is little known.

For the first and last time in his life German General Erich von Falkenhayn had an imaginative plan. His novel vision was to make Verdun a very special killing-ground for Frenchmen. His intention was to sucker the enemy into a protracted battle of attrition, to make a surgical incision upon the body of France through which it would bleed and bleed and bleed until dead.[104] He had in fact invented an interesting variant of the "luring defensive" that a later generation of French military theorists would call an *abcès de fixation*.

And all went well at first. When Falkenhayn's attack went in on February 21st the French were taken quite by surprise, strategically by the place of his main attack and tactically by its time and strength. French losses in the first week of surprise were nearly *three times* those of the Germans even though the latter were on the offensive, which doctrine had decreed was more costly for the attacker than the defender. Falkenhayn had flaunted and disproved this doctrine. But this valuable lesson in the economical consequences of surprise-through-deception was forgotten in the holocaust that followed. What had gone wrong?

---

[104] General von Falkenhayn, *The German General Staff and Its Decisions, 1914-1916* (New York: Dodd, Mead, 1920), 249-250, 255-273.

To prevent leaks of his intentions, Falkenhayn had severely limited the number of knowers by tight security and partial disclosures. Senior headquarters were not informed; parallel headquarters were prevented from discovering the secret by the simple technique of excluding their liaison officers from the Fifth Army front at Verdun; and even Falkenhayn's own artillery adviser was not informed until it was too late to appropriately modify the artillery program. Even Germany's Austrian allies were not told—to their subsequent annoyance.[105]

Falkenhayn had brought on catastrophe by his own secretiveness. By neglecting to inform his superiors and colleagues that his intention was a *strategy of attrition*, they soon joined the French in viewing Verdun as a symbol, a precious trophy, that must be held at any cost. Thus did that smashed piece of real-estate become a killing ground for the Germans as well. When the battle ended 10 months later on the last day of 1916, the total casualty rates only slightly favored the Germans (337,000 Germans to 377,000 Frenchmen, or 1-to1.1). Moreover the bankruptcy of the German strategy at Verdun is clear from the statistics for the other sectors of the Western Front during the same period, statistics that favored the Germans 1-to-1.5. In other words, the existing German strategy was already almost half again more effective for attrition than Verdun itself.[106]

CASE 2.4.2:
Dienbienphu, a Disastrous Lure, French Indochina 1953-54

General Henri Navarre very deliberately committed major elements of his French Army in Indochina to its long agony at Dienbienphu. His intention was that it serve as an *abcès de fixation* to lure Vietnamese General Vo Nguyen Giap's regular divisions into a set-piece battle on a killing-ground of Navarre's choosing. To induce Giap to adapt a frontal attack or siege tactic, the bait had to appear temptingly weak while being, in fact, strong. This type of baited trap—on a smaller scale—had been and continued to be a standard tactic among the French.[107]

The French had taken the isolated mountain top prize of Dienbienphu from the Viet Minh by airdrop on 20 November 1953 and gradually increased its garrison to a peak strength of 13,000. General Giap accepted the challenge and mounted a siege, bringing in more and more troops, just as Navarre had wanted. With complete command of the air, Navarre assumed he could reinforce and resupply as long as necessary. The early Viet Minh counterattacks were easily

---

105 Falkenhayn (1920), 255, 264-265.
106 Whaley, *Stratagem* (1969/2007), Example B8a (Verdun).
107 Bernard B. Fall, *Hell in a Very Small Place: The Siege of Dien Bien Phu* (Philadelphia: Lippincott, 1967), 5, 49.

repulsed with high casualty ratios (well over 3-to-1) for the Viet Minh. All was going according to Navarre's plan.

Then something began to happen that Navarre had not anticipated. Dienbienphu was becoming a household word not just in France but throughout the world. This minor bit of real estate had become transformed into a symbol of prestige and honor for France—and for Navarre.

Everything was now going according to Giap's plan. Instead of merely trickling in one unit after another to be destroyed piecemeal, he managed to gradually build up the besieging force to 50,000 combat troops and 55,000 support troops. He had won the race of men and matériel. The French had been able to commit only 9 battalions of infantry, insufficient even for a tight perimeter defense with limited air support, while the Viet Minh had moved up 3 divisions, supported by massed artillery. Thus Giap was able to bring overwhelming pressure on the French hedgehog defense, once the latter had reached the limit of their ability to reinforce it.

Although Navarre was fully aware of his own force limitations, he had grossly underestimated not only his enemy's strength but both their speed of reinforcement and their ability to stand and fight despite complete French air control. And, as the defense perimeter was squeezed tighter and tighter, the helipads were becoming vulnerable to artillery fire. Air resupply continued but was limited to airdrops. Evacuation, much less reinforcement, had become impossible. The French were now caught in their own trap.

Dienbienphu was overrun on 7 May 1954. The French could not even claim that the enemy had won only a Pyrrhic victory, as losses had been nearly equal (18,916 French to 22,900 Viet) in a war where it was the French who were chronically short of manpower. Navarre could (and repeatedly did) point out that only 4% of his total force had for half a year held 20% of the enemy's, including 60% of their main combat force. However, important as this undoubtedly was in its effect on the outcome of many minor operations during the six-month siege, it begs the fact that Navarre had lost the one politically decisive battle of the entire war.

The Vietnamese history professor turned victorious general summed up:[108]

> The French Expeditionary Corps faced a strategic surprise—it believed that we would not attack and we did attack; and with a tactical surprise—we had solved the problems of closing in, of positioning our artillery, and of getting our supplies through.

Giap's opening clause is most interesting. It indicates he had indeed taken Navarre's bait and was lured into an attack. However, as his second clause

---

108 General Giap as quoted by Fall (1967), 51.

shows, he perceived a different opportunity to trap the French. Rather than play by Navarre's rules, he changed the rules. He made his own rules and won.

On his part, Navarre, victim of his own hubris, had grown so fascinated by his scheme that he missed the opportune moment to liquidate it and safely evacuate the garrison. It was later than he thought.

■ ■ ■ ■ ■

The great lesson of Verdun and Dienbienphu is that the perpetrator must avoid becoming so enthralled with his own lure that he forgets its original purpose as a bogus payoff and becomes almost literally as well as figuratively hoist on his own petard. Fortunately this worst-case situation will arise only in high visibility operations that interface between war and politics where the commander and his deception team can be superceded by meddlesome political generals and politicians.

2.5. Aborts

A minor and peripheral part of the failure question is that of deception plans that abort. The topic is interesting only because it has never been discussed. It is important only because of the lessons learned. First a few examples:

CASE 2.5.1:
An "A" Force abort, Cairo 1942

The Black Code was U.S. State Department code in use at the outbreak of WW2 by all American military attaches throughout the world. In August 1941 an Italian employee in the U.S. military attache's office in Rome photographed a copy. Although Italian Military Intelligence gave its German counterpart, the Abwehr, selected decrypts, they did not—as widely published—pass over the code itself. However, the German OKW's Cipher Branch managed to solve the Black Code on its own; and from late September the Germans were reading the American military attaches' reports. It was soon evident that those of Colonel Bonner F. Fellers were of great value. Fellers was the military attache in Cairo and was sending to MILID WASH (the Military Intelligence Division in Washington) frequent, accurate, and detailed reports on the British Army in the Middle East—its deployments, reinforcements, and state of equipment, as well as unusually perceptive assessments of their moral, state of training, and even current plans & operations. At Rommel's headquarters these became known as "the Good Source" and Rommel himself referred to them affectionately as his "little fellers". Even Hitler joked about these precious gifts from the American Embassy in Cairo.[109]

This leak continued for over 10 months until the pre-dawn of July 10th when a large Australian raid penetrated the Italian line, overran Rommel's radio intercept company, and brought back its records. British Intelligence was appalled to discover the degree of its radio insecurity, which they promptly tightened. And, now armed with full details of German intercept SOP, the British now used it to work better radio deception on Rommel's new and inexperienced German intercept company when it arrived.[110]

These captured documents also revealed that the enemy was reading Feller's reports, treating him as an "unconscious source of proven accuracy". Therefore, Lieut.Col. Noël Wild, Clarkes deputy at "A" Force urged that Feller's dispatches to Washington be continued in the Black Code with suitably doctored intelligence. However, and contrary to several published accounts, this was not done, as Wild himself confirmed. Wild's plan was rejected because the senior political officials who had to be consulted refused to risk offending MILID WASH by using one of its attaches as a stooge. The Americans were informed, they dropped their Black Code, and Fellers was recalled.[111] Wild's promising deception plan had been aborted by political, specifically diplomatic considerations.

CASE 2.5.2:
Aborts by Twenty Committee, London 1941

London's Twenty Committee had very mixed results in its first full year of operation. Its few partially successful operations (such as Plan IV and Plan MIDAS) were outnumbered by the aborts (such as Plan STIFF and Plan PAPRIKA), disappointments (such as Plan MACHIAVELLI and Plan PEPPER), and outright failures (such as Plan I and Plan MICAWBER). The main cause was sheer lack of coordinated planning.[112]

CASE 2.5.3:
The "Italian King's Proclamation" abort, London 1943

Someone suggested that the British Psychological Warfare Executive (PWE) might assist the invasion of Sicily by broadcasting a bogus proclamation by the King of Italy.

---

109 David Kahn, *The Codebreakers* (New York: Macmillan, 1967), 472-477; David Kahn, *Hitler's Spies* (New York: Macmillan, 1978), 193-195.
110 Sefton Delmer, *The Counterfeit Spy* (New York: Harper & Row Publishers, 1971), 29-30; Brown (1975), 104.
111 Delmer (1971), 30-31; Mure (1980), 120-121.
112 Masterman (1972), 82-89; Holt (2004), 151, 828, 832, 837.

Colonel Bevan, the Controller of Deception, proposed a leaflet saying the King had asked for an armistice, but objections were made that this would discredit the Allied propaganda machine. However, both Eisenhower and Alexander supported Bevan's proposal as a way of reducing Italian resistance in the crucial early days of the invasion. Accordingly a text was drafted that had the "King" saying he had dismissed Mussolini and agreed to an armistice to save the country from anarchy.

Initially Churchill and the British and U.S. Chiefs of Staff agreed to this ruse. However, second thoughts prevailed; so they decided it would quickly backfire, showing the world that Allied propaganda had sunk to the level of that of the Axis. Consequently the false proclamation plan was cancelled and Mussolini remained in office a few more weeks.[113]

CASE 2.5.4:
Plan JAEL, London 1943

Plan JAEL followed the fiasco of COCKADE (Case 2.2.6) and was its obvious attempt at a successful replication. However, beginning in early fall of 1943, Col. Bevan and his LCS deception planners in London initially repeated COCKADE's error of trying to simulate threats against Hitler's Fortress Europa at too many places, from too many directions.

Fortunately Col. Dudley Clarke intervened from Cairo. He convinced London of this flaw but London's fiddling adjustments gradually led to JAEL simply vanishing to become replaced in January 1944 by BODYGUARD, the comprehensive and tightly focused new plan that, together with its superb cross-Channel FORTITUDE, would produce an enormous and decisive success for deception.[114]

CASE 2.5.5:
Aborts by the 23rd Hq Special Troops, France 1944-45

In their support of the U.S. Army ground forces in France in 1944-45, the 23rd Headquarters Special Troops deception unit worked up a total of 26 plans, 21 of which became actual operations and 5 of which were aborted (see Case 2.1.4). Of the 5 aborts, 2 were caused by last minute changes in Allied plans, 1 occurred on the beaches of Normandy where the deception troops found the situation different than expected, 1 occurred because the unit arrived too late to be of any use, and the last because of administrative foul-ups including the embarrassing theft of the 23rd's liaison officer's van with needed materials.[115]

---

113 Cruickshank (1979), 57-58.
114 Holt (2004), 496-498, 506, 507, 825; Mure (1980), 237.
115 *Official History of the 23rd Headquarters Special Troops* (1945).

■ ■ ■ ■ ■

The most common cause of aborts occurs when higher commands change the priorities among various theaters or sectors of combat operations. A carefully planned operation and its associated deception plan on one front can get cancelled because promised reinforcements or supplies get shifted to another front. Or the Commander's own sector may be unexpectedly stripped of key units by his superiors who see a greater need for them elsewhere. This applies at all levels, from those of minor tactics to grand strategy. An example follows:

CASE 2.5.6:
Aborts by "D" Division, Burma 1942-45

Allied WW II grand strategy was set in Washington and London. First priority went to the European Theater, second to the Pacific, and third—a poor third—to the China-Burma India Theater. Consequently the battle plans of the CBI Commander—first Wavell then Mountbatten—were plagued throughout by chronic uncertainty over the amount and timing of reinforcements.

The CBI Commander's deception was handled by "D" Division, headed by Col. Peter Fleming. His biographer describes the dilemma:[116]

> Time and again Mountbatten projected a major assault, only to have the promise of essential equipment (such as landing-craft) withdrawn at the last moment, so that all his plans had to be changed; and time and time again Peter, with his uncanny knack of anticipating the future, had produced as his deceptive feint the very plan on which the Supreme Commander then had to fall back. As Peter remarked to one of his colleagues, it was impossible to tell a convincing lie unless what he knew what the truth was going to be, and in the end the number of occasions on which he involuntarily foreshadowed the truth proved rather embarrassing.

This lack of long-term plans, combined with the inefficiency of the Japanese themselves, ensured that the deception practiced in South-East Asia could never be anything as sophisticated as that carried out from London or from Cairo.

■ ■ ■ ■ ■

---

116 Hart-Davis (1974), 284.

One specialized branch of deception occurs during R&D when it is used to induce misperceptions about the value of scientific or technological breakthroughs, such as one's acquisition of some new weapon or sensor system. The consequence of such pay-off misperception is often a correspondingly dysfunctional response, a wasteful counter-effort. A successful example is the British pretense of having developed an infrared detector (CASE 2.2.4). Two examples of failure follow:

CASE 2.5.7:
Plan RAINBOW, CIA, 1956

An interesting historical example of R&D deception failure with some present relevance was Plan RAINBOW. This was a CIA deception plan in the late 1950s to simulate an Anglo-American scientific-technological breakthrough. RAINBOW intended to convince the Arab oil-producing countries, who even then were engaging in a bit of oil blackmail, that the USA was on the verge of developing a revolutionary new energy source. The CIA's expectation was that if this fiction was believed, the Arabs would underestimate American dependence on oil, giving American diplomats stronger leverage in Middle East negotiations.

The plan was aborted when the CIA learned of independent technical reports—presumably known or soon-to-become-known to the Arab oil producers—that proved such a source, even if theoretically possible, could not provide enough new energy to significantly cut USA dependence on Arabian oil.[117]

Case 2.5.8:
R&D Deception Aborts in the Czechoslovak Disinformation Department, 1960s

Every week or so someone in the Czechoslovak Interior Ministry would put forward a new suggestion about how to fool Western intelligence services about Research & Development (R&D) behind the Iron Curtain. All such suggestions were given the most careful attention by the Disinformation Department. All were highly imaginative, most were rejected as too expensive or too risky, and only a few showed real promise.

However, even the most promising proposals usually foundered because of the refusal of senior Czechoslovak or Soviet scientists to risk their reputations by contributing their names to an operation they rightly felt would eventually get exposed as a hoax. And it was only their top men of science that the Russians or Czechs felt they could effectively use. Notional scientists would not do

---

[117] Miles Copeland, *Without Cloak or Dagger* (New York: Simon and Schuster, 1974), 192-195.

because the Western intelligence services would probably be able to learn that they were mere phantoms. Even the few junior scientists who were willing to cooperate would not do either, because they would lack credibility in the West without the imprimatur of senior colleagues or simply because they would be highly suspect.[118]

■ ■ ■ ■ ■

Deception plans abort for the same reasons any plans abort: circumstances change, rendering the original plan obsolete. These circumstances may be, as with Plan RAINBOW, new knowledge. It may result from personnel changes among the planners. Most often, however, the Commander—either ours or theirs—merely changes his mind about what he intends to do. Or has it changed for him, as when the supreme command suddenly removed the left-flank divisions that Alexander had intended to use for his offensive in Italy in summer 1944. Similarly, "A" Force's deception plans aborted the two times when Rommel preempted the British offensives with ones of his own.[119]

Deception aborts are inconsequential if they are still in the planning stage—at worst a real operation allowed to proceed without cover; at best a useful training exercise with lessons learned. I am unaware of any abort that caused serious harm. But once launched against the enemy, aborted deception operations are apt to cause great harm as Hitler did in his summer 1942 offensive in Russia.

2.6. Pseudo Failure

This chapter warrants a brief look at the problem of pseudo failures—those cases that some writers have labeled total or near-total failures, which in fact were successes, at least in large part. Three factors account for these frequent false accusations of deception failure. First, too many military writers make the fundamental error of failing to "look at the other side of the hill" to see how the victim perceived the deceiver's action. Second, and at least equally important, is the general tendency of military writers to rate all cases of only partial failure as if they were total failures. In fact, while only a small proportion of deception operations meet or exceed their designer's expectations, only a very small proportion completely fail much less backfire. Third, and most important, is that most military writers undervalue the cost-effectiveness of deception. They resist the notion that deception is a fundamentally simple process. Every master magician and con artist knows better.

---

118 BW interview with Maj. Dr. Ladislav Bittman, 1972.
119 Whaley, *Stratagem* (1969/2007), Cases A26 and B22.

For example, we saw (Case 2.1.4) that the official historian of the 23rd Headquarters Special Troops underestimated the effects of at least two of their 21 deception operations because of inadequate feedback. Let's examine three other instructive examples.

CASE 2.6.1:
German Operation ALBION (= SHARK + HARPOON) as Cover for the Invasion of Russia 1941

At least four studies have written off Hitler's Operation ALBION as a failure or near failure.[120] They are far too pessimistic.

On the contrary, ALBION and its two components, SHARK & HARPOON, was more than half successful in reaching its dual goal. First, it succeeded in maintaining Britain's expectation of an imminent German invasion until well into 1941. Second, simultaneously and more importantly, it was an important factor in lulling Soviet Intelligence, specifically Stalin, into failing to appreciate how close German was to invading Russia.[121]

CASE 2.6.2:
Operation COPPERHEAD, Gibraltar & Algiers 1944
Did the touring actor play to an empty house?

The famous "Monty's Double" ruse played to mixed reviews. A highly romantic version that claimed success was circulated after the war in a ghostwritten memoir of the principal actor[122] and in a movie based loosely on the book, both being titled "I Was Monty's Double". We also have three published accounts: a memoir by the case officer[123] and interviews with two of the planners.[124]

Plan COPPERHEAD originated with a coincidence on 14 March 1944. That day the Deputy Chief of the Committee of Special Means (GSM), Lieutenant-Colonel John V. B. Jervis-Reid, noticed a photograph in a London newspaper. At first glance it looked like a picture of General Montgomery, but the caption read "You're Wrong—His Name is James". Monty's look-alike was Lieutenant Meyrick Edward Clifton James of the Royal Army Pay Corps and a peace-time minor actor in British provincial theater. The photo showed James playing Monty in an armed services show at the Comedy Theatre in London.

---

120 Alan F. Wilt, "'Shark' and 'Harpoon': German Cover Operations Against Great Britain In 1941," *Military Affairs*, Vol.38, No.1 (Feb 1974), 1-4; Maxim (1982), 17-19, plus additional information from Dr. Wilt; Cruickshank (979), 207-210; and *FM 90-2* (1988), Chapter 7.
121 Whaley, *Codeword BARBAROSSA* (1973), 172-173, and throughout.
122 Clifton James, *I Was Monty's Double* (London: Rider, 1954). The popular movie version was released in 1958.
123 Stephen Watts, "I Was Monty's Double Twice Removed", in the author's collection of memoirs, *Moonlight on a Lake in Bond Street* (London: The Bodley Head, 1961), 158-173.
124 Interview with J. V. B. Jervis-Reid in Brown (1975), 608-612. Interview with Noël Wild in Mure (1980), 251-252.

This photo gave Jervis-Reid an idea. Why not have James impersonate Monty, sending him off on an "official" tour of military posts in the Mediterranean to suggest a false scenario to German Intelligence, namely that the appearance of the commander of the British 21st Army in the Eastern Mediterranean at this time would reinforce the parts of the overall BODYGUARD deception designed to indicate an early invasion of southern France.[125]

Major Stephen Watts became case officer, coaching James for his most important role and coordinating with Col. Wild (head of CSM) and Col. Michael Crichton (the "A" Force representative at General Eisenhower's HQ in Algiers) who would join them on arrival. James seemed perfect for the role in face, physique, speech, and gesture. If only he could control his taste for booze and cigars—two popular pleasures that the puritanical Monty was publicly known to abhor. Then, on May 26th, off on the high visibility tour—first stop Gibraltar where showtime was breakfast at the airport.

The first omen of potential failure occurred during the flight. Overawed by the importance of his role, James got stage fright and overcame this by sneaking back to the converted Liberator's W.C. and getting soused on his smuggled hip-flask of gin. Drastic measures by Wild and Watts sobered him up enough to put in a shaky but credible performance at the airport and that evening as guest of honor at the Governor's dinner party where "Monty" babbled about a mysterious "Plan 303". The following day he flew to a simulated high-level conference in Algiers.

That evening James wandered away from Major Watts who found him walking about the streets of Algiers, drunk and smoking a cigar. Operation COPPERHEAD was promptly cancelled and James quietly flown back to England and anonymity. Anthony Cave Brown asserts that no evidence emerged then or after the war that showed the Abwehr had taken any notice of Monty's Double. Nothing gained, nothing lost, he concludes.[126] The truth is more complex.

In fact, ULTRA revealed quite soon during the operation itself that the Abwehr had indeed reported Monty's presence. This was the real reason COPPERHEAD was ended so abruptly—it had succeeded. There was no reason to prolong the risky charade. Col. Crichton got his "General" a private room at the St. George Hotel in Algiers, set out a bottle of whisky, a syphon, a couple of glasses, and invited James to join him in a drink. Excusing himself for awhile, Crichton returned to find an empty bottle and a snoring actor. Indeed, instead of being returned to London in disgrace, as Anthony Cave Brown

---

125 Mure (1980), 126.
126 Brown (1975), 612.

claimed, next morning Crichton rewarded his hungover actor with a week's leave incognito in Cairo on—as Monty had insisted—full General's pay.[127]

While James caught the blame, the plan may have been slightly flawed from the outset. The British placed great store on Montgomery. Churchill, the British press, and Monty's own overworked press officer had built up his image in Britain and America. They assumed German Intelligence would be equally awed by all this hoopla. I have never seen any evidence on the German side that this was true, although the Abwehr did closely and nervously track the movements of General Patton, whom they had learned to respect and fear.

A fair conclusion is that COPPERHEAD was successful in providing at least one more piece of "evidence" for the Abwehr to fit into their jigsaw puzzle, although a somewhat smaller piece than its instigators had intended.

CASE 2.6.3:
Operation ERROR as a Partial Success, Burma 1942

Soon after Colonel Peter Fleming arrived in Delhi in April 1942 to became General Wavell's Director of Deception, Wavell outlined a plan after dinner with his ADC Bernard Fergusson and Fleming. Wavell told his guests that he wanted to revive "the Meinertzhagen ploy" and retold the famous exploit.[128] (Case 5)

Accordingly, Fleming prepared a set of fake documents that indicated major reinforcements were due soon. His intention was to deter the Japanese from invading India by making them believe that Wavell's defenses were not as weak as they were. These documents were placed in Wavell's own dispatch case, which in turn was left in a staff car that Fleming deliberately crashed on May 3rd on a road in the path of the Japanese advance. Indications were left at the scene that General Wavell had himself been injured in the accident.[129]

British military historian Ronald Lewin states that, "Neither at the time, however, nor as a result of post-war investigations and interrogation of Japanese officers, did it ever become clear whether this Operation *Error* had achieved its purpose." Lewin concedes only that, "Still, it was a useful exercise. At least it indicated that Wavell was not stale."[130] And, Lewin might have added, that it had certainly been a useful introductory exercise in deception planning & ops for Fleming.

But Lewin was simply wrong in his conclusion. Had he read further in his own cited source he would have noticed that Fleming "did in 1944 get from Chinese

---

127 Mure (1980), 251-252.
128 Hart-Davis (1974), 265-266.
129 Hart-Davis (1974), 266-269; Lewin (1980), 174.
130 Lewin (1980), 174.

intelligence sources a gratifying report that during the first Burma campaign the Japanese had captured important documents which indicated that India's defensive potential was greater than Tokyo had supposed."[131] Thus it seems that the Wavell-Fleming melodrama had played well in Tokyo. However, this is a rare and interesting example of a deception that while "successful" in getting the enemy to believe the false intelligence was, in fact, irrelevant and unnecessary—ironically, the Japanese had no intention of over-extending themselves by invading India.

ERROR holds a final lesson that deserves a section to itself. Hence the next remarks and cases on the neglected topic of the *potential* for failure inherent in every deception plan.

2.7. Potential Failure

Every deception carries with it the seeds of its own failure. Leave aside the ill-conceived plans, the target's inattention, inadequate deception assets and channels, and outright bungles. These have been covered above under the previous four circumstances of deception failure. Here we consider the fifth, the ever-present possibility of detection.

Deceptions—all deceptions—are detected when the target spots and understands the incongruities inherent in every deception. There will be at least two—a discrepancy in the attempt to hide the real thing (object or idea) and another in the attempt to show something false in its place.[132]

The great majority of historians, myself included, normally avoid asking "What if?" We consider most "if in history games" pointless and leave such speculations to writers of historical and science fiction with their "alternative universes". We stand content with what actually happened—Wolfe reached the Heights of Abraham by an indirect route, surprised the French, and captured Quebec. It seems futile to ask "What if Wolfe had made a different choice?" But there is an exception to this rule. It occurs whenever the historian intervenes to criticize the actor's plans or operations. This then becomes a hypothetical exercise seeking "lessons learned" in the hope that we can profit from the mistakes of others.

---

131 Hart-Davis (1974), 269.
132 Whaley & Busby (2002); and Whaley, The Maverick Detective: The Whole Art of Detection (manuscript in progress since 1986).

CASE 2.7.1:
Operation ERROR as a Potential Failure, Burma 1942

In the preceding case we saw that Col. Fleming's Operation ERROR was a marginal success. Let's examine where it could easily have gone wrong. Moreover we are fortunate in having the ruthless critique of ERROR'S own planners, Fleming and Wavell.

In his post-operation report, Fleming posed the question, "Did the enemy see through the documents?" and answered himself as follows:[133]

> More than 80 per cent of the documents, personal and official, were genuine; the important minority of fakes were carefully concocted and, being mostly couched in allusive or otherwise slightly indefinite terms, were almost, if not quite, proof against definitive exposure. In other words the forgeries, though they included suggestions which my have been disbelieved by the enemy, contained no statement of fact which he can *disprove*.

Then, in answer to the follow-up question "Did the enemy see through the ruse itself?", Fleming wrote:[134]

> Sherlock Holmes, had he been promptly on the spot, could without difficulty have deduced from the skidmarks and other indications that there was something fishy about the supposed accident. But it is safe to assume that, long before his Japanese equivalent could have reached the Ava Bridge, large numbers of Dr Watsons had decisively prejudiced all chances of reconstructing the crime....
>
> [The] worst thing we could have come up against would be a German liaison officer at Japanese HQ in Burma who was familiar with the story of Meinertzhagen's haversack. A great deal will depend on at what stage and on what level suspicions are first aroused.

On this last point, General Wavell penned his own confession:[135]

> I always realized this danger, and after Fleming had gone thought I had made a bad mistake in including in the exhibits a paper which connected me with the writing of Allenby's life [with its prominent account of the haversack ruse], but it was too late to recall it.... Any harm done if the Jap decides it was a plant? I don't think so.

---

133 Hart-Davis (1974), 268.
134 Hart-Davis (1974), 268-269.
135 Hart-Davis (1974), 269.

These comments show healthy self-criticism. Let's look at a contrary case, one where the planners were too uncritical of their scheme.

CASE 2.7.2:
TRICYCLE-ARTIST-GARBO as a Potential Blown Network, Britain 1944

The success story of three Twenty Committee double agents, TRICYCLE, ARTIST, and GARBO has been detailed in Sefton Delmer's unauthorized leak, Sir John Masterman's and Tomás Harris's official histories, and in TRICYCLE's and GARBO's own memoirs.[136] Yet how close the Committee came to blowing the entire FORTITUDE deception plan. From "A" Force perspective in Cairo—where double-agents had been used for deception with consistent success—Major Mure analyzes that potential downside part of the story:[137]

> Even by 1944, the lesson had not been learned in England that, since deception was an essential part of the operational plan, double-agents used for this purpose must never be allowed to meet or come within the power of their employers....
>
> It was our old friend Tricycle and his Abwehr friend Jebson who nearly brought the whole deception campaign to disaster. Not only was Tricycle put forward to the XX Committee as a suitable deception agent but he was permitted to continue his travels between Spain and London to contact his controllers. At the same time, his friend Jebson ... was recruited as a British agent ... and left in possession of up-to-date knowledge of *Tricycle's* deception activities.
>
> At the end of April—a month before the opening of the Second Front, Artist was arrested by the Gestapo and it was pound to a penny that, under torture, he would blow *Tricycle's* role in the deception plan. This near disaster illustrates the continuing and fatal confusion of penetration with deception.

Mure's point is well taken. Just because German counterintelligence was too weak to unmask these British double agents is no excuse for assigning both deception and intelligence roles to individual double agents. I understand the temptation to get two products from one agent. But absolute compartmentation

---

136 Sefton Delmer, *The Counterfeit Spy* (New York: Harper & Row Publishers, 1971); Masterman (1972); Tomás Harris, *GARBO: The Spy Who Saved D-Day* (Richmond: Public Record Office, 2000); Dusko Popov, *Spy/Counterspy* (New York: Grosset & Dunlap, 1974); and Juan Pujol with Nigel West [pen name of Rupert Allason], *Garbo* (London: Weidenfeld and Nicolson, 1985). For Mure's controversial criticism see Holt (2004), index; and Terry Crowdy, *Deceiving Hitler: Double Cross and Deception in World War II* (Oxford: Osprey, 2008), index.

137 Mure (1980), 248-249.

between deception and intelligence assets should be mandatory. Why? Because, if your opponent captures one of your intelligence agents all you lose is just that one source of intelligence. But, if your opponent gets one of your deception agents, you also risk disclosure of your real military or political plans and intentions.

3. Coping in Military Practice

How do Commanders and their planners deal with deception failure, both potential and actual? First let's explode two of the many myths surrounding and inhibiting the practice of deception.

3.1. Soldier's Myths

One influential myth is that one must never use the same ruse twice. For example, the reader might conclude that the old "Haversack Ruse" (long falsely claimed by Major Richard Meinertzhagen) has been done to death. This would be a mistake. It had partly worked for Maj. Meinertzhagen in Palestine in 1917; for Col. Conger at St-Mihiel in 1918; for Col. De Guingand at Alam Haifa in 1942; for Lieut.-Col. Fleming in Burma in 1942; for Lt.-Cdr. Montagu in Spain in 1943; probably for British M.I.5 in the later stages of the CICERO game in 1944, and it worked for the Germans in the Bavarian Redoubt in 1945. In all, I have traced 40 Haversack Ruse type cases, 35 involving the planting of false documents and 5 involving unintended loss of real ones. Omitting the 10 planting operations that were aborted, leaves a fairly cost-effective 60% (15 of 25) completed efforts to float false plans enjoyed at least some success. Conversely, of the 5 real battle plans, 4 were disbelieved and the other had an unknown effect.[138]

In fact, no ruse can be made obsolete through repetition. The worst that can happen is that the enemy will sooner or later recognize this as a preferred type of ruse for the deceiver and examine all such cases with heightened suspicion. But, even then, the enemy is forced to decide each time whether the documents are real or fake, whether the attack is real or a feint, whether the tanks are real or dummies, whether the "Field Marshal" is real or an impersonator, and so forth. Moreover, a wily deceiver might even use stale ruses as "smoke" to distract enemy Intelligence from more subtle ruses.

The second myth portrays deception as a dangerous game that should be avoided. Some academic military specialists, assuming that the chance of deception failure is so high and the costs so great when it happens, conclude that it is better to have no deception at all. This is overly pessimistic advice. Fortunately for their careers, few military commanders listen to this particular piece of advice from the experts. In one study I found that even in those

---

138 Based on the 40 cases analyzed in my "Partial Inventory of Real Plans Lost & False Plans Planted" in Whaley, *Meinertzhagen's Haversack Exposed: The Consequences for Counterdeception Analysis* (FDDC, May 2007), 25-29. This data replaces that in Whaley, *Stratagem* (1969), 230, which had reached a false conclusion based on a small sample of only 10 cases.

few cases (10 out of 84 attempts) where deception apparently failed totally, the average results of the ensuing battles were no worse than for those cases (93) where deception was not even attempted. Indeed, paradoxically, they were slightly better.[139] My hunch is that just enough deception-induced misperception, too small for the victim to notice, was getting through to cause a slightly unbalanced response.

Neither the soldier's manuals nor the general's' war plans tell them what to do when deception fails. The usual attitude is "Well, we tried." Because, as noted elsewhere,[140] deception seldom costs much; most commanders just write a failed deception off as a loss and either abort the real operation or, more likely, charge on in without the advantage of surprise. In either case, the logic is that a failed deception is no worse than not having tried at all. I agree with the logic but would hope for a more creative solution. Three soldiers—all "A" Force deception planners—are known to have found such a solution.

3.2. Soldiers' Practice

I have found only three cases where military officers overcame this self-crippling sense of helplessness and redesigned their failed deception plans to take positive advantage of the new situation. The first two cases are rare examples of total failure of the initial simulative camouflage when the enemy detected it, but then turned to advantage.

CASE 3.2.1:
Col. Clarke and the Aircraft Struts, Egypt 1942

Sometime around mid-1942 Colonel Dudley Clarke, head of "A" Force, the Cairo-based British deception team, was informed that the Germans had learned to distinguish the dummy British aircraft from the real ones because the flimsy dummies were supported by struts under their wings. Here was a camouflage deception that had completely failed.

A commander with a straightforward mind, having recognized this telltale flaw in the dummies, would have ordered his camouflage department to correct it. But Clarke's devious mind immediately saw a way to capitalize on the failure. His instant order was "to put struts under the wings of all the real ones, of course!"[141]

By putting dummy struts on the real planes while grounded, enemy pilots would avoid them as targets for strafing and bombing, Moreover, it would

---

139 As recalculated from tables in Whaley, *Stratagem* (1969), 163, 199.
140 Whaley, *Stratagem* (1969), 232-244.
141 David Mure, *Master of Deception* (London: William Kimber, 1580), 98.

cause the German photo-interpreters to underestimate and mislocate the real RAF planes.

Fortunately, unlike Ralph Ingersoll when he invented the brilliant "Two Patton's Ruse",[142] Clarke did not dismiss his innovation with the struts as a merely clever one-off solution to a specific problem. It is interesting that Clarke had effectively invented the double-bluff as a general solution to a general problem, an important one. And he taught his staff this new category of military deception, as we shall see in the next case.

CASE 3.2.2:
"A" Force and the Dummy Tanks, Egypt 1942

During the retreat of British Eighth Army to Alamein in June 1942 its 4th Armoured Brigade with over 120 American "Stuart" light tanks had been so heavily mangled by Afrika Korps that it ceased to exist as a fighting unit and its 10 surviving tanks were absorbed by another unit. "A" Force promptly took advantage of this by adding the defunct 4th to its growing notional order of battle. Commander of the now bogus 4th Armoured Brigade was "Brigadier" (actually Major) Victor Jones. His task was to maintain a counterfeit threat to Rommel's southern flank. Soon, however, German military intelligence detected the sham, reporting that the tanks of the 4th Armoured were plywood and cardboard with a gadget at the rear to simulate tank tracks. This report was sent by radio and therefore duly read by the British ULTRA team in London, which passed it along to "A" Force in Cairo.[143]

"A" Force took prompt advantage of this by having the 4th Armoured became a 50-50 mix of dummy and real tanks. Consequently, when Rommel moved against Montgomery's southern flank during the Second Battle of Alam Halfa on August 30th his assault unit, its supporting tanks well to the rear, was surprised to meet real gunfire from British light tanks that they had expected would be only dummies.[144] Indeed it was the action in this sector that slowed Rommel's attack to such a degree that his whole battle plan began to unravel.[145]

"Brigadier" Jones and his 4th Armoured Brigade then served in a suitable supporting role for the overall deception plan (Plan BERTRAM) for the upcoming Battle of Alamein. After Alamain Jones was once again a major.[146]

---

142 Whaley, *Turnabout: Crafting the Double-Cross* (FDDC, Jan 2010), Case 5.4 (pp.69-71).
143 Mure (1980), 126.
144 Mure (1980), 97, 126-127, 128.
145 W.G.F. Jackson, *The Battle for North Africa, 1940-43* (New York: Mason/Charter, 1975), 340-341.
146 Mure (1980), 141.

Case 3.2.3:
Gen. Alexander's deception team double-bluffs, Italy 1944[147]

This last example of a double-bluff is an audacious and risky one, a last minute improvisation forced upon the deception planners by sheer urgency.

Operation OLIVE was General Sir Harold Alexander's decisive offensive to break the German line across central Italy in summer 1944. In early June, he planned his main attack with his American and French divisions to go through the mountainous center, with a feint by his British divisions along the eastern flank on the Adriatic. On July 5th Alexander learned that London and Washington had decided to strip him of many of his American and all his French divisions, sending them away for the amphibious invasion of South France. Consequently, on August 4th, he reluctantly decided to make the real attack up the Adriatic coast with his British troops, leaving the denuded American force to conduct the feint at the center. In other words, the very strategy that the current deception operations were communicating successfully to the Germans.

This reversal of strategy required a plausibly readjusted deception. Moreover, with OLIVE D-day set for August 25th, Alexander's deception planners[148] had only three weeks to reverse the enemy's perceptions—a feat that military history shows to be very rare.

The essence of the new deception plan was to work a double bluff by having the Germans now believe that the old "evidence" fed them had been and still was the Allied deception operation. This was done by a new and elaborate program of radio deception. As far as I am aware, this is the only case where a military deception operation was *deliberately* turned back on itself—a stratagem to attempt to discredit earlier disinformation by exposing it for the deception it was.

The new deception was successful enough to gain surprise of place and strength and of timing as well—two German divisions were uselessly tied up in reserve for a "possible" attack at the center, one division at the point of real attack was caught being relieved, and both the field commander and a key divisional commander were still off on leave.

---

147 For details and documentation see Whaley, *Stratagem* (1969/2007), Example B38 (Battle of the Gothic Line).

148 The deception planners at Alexander's HQ in Italy were a small detachment from "A" Force headed by Lieut.-Col. Michael Crichton. Holt (2004), 610, 625-628; Mure (1980). 259.

PART TWO:
Solutions to Failure

The three previous chapters have shown the various ways in which deceptions can fail. The final three chapters will describe practical solutions to deception failure.

4. Coping in Theory

Here, at last, we must face the central question raised in this paper: What can deception planners do when a plan fails? What steps can they take to minimize this happening and, if it does, to turn failure into an, at least, limited success.

4.1. Toward Theory

Unique examples should always be of keen interest to the theorist because they reveal gaps in existing theory. Being interested in the consequences of failed deceptions, I was immediately struck when back in 1969 I stumbled across the dilemma faced by British General Alexander at the Gothic Line in Italy (Case 3.2.3). However, I interpreted it too narrowly, seeing only a unique and therefore fascinating case of deliberately blowing one's own deception plan.

Eleven years later, when I first read the Clarke "struts" ploy (Case 3.2.1), I suddenly recognized that these two examples sounded a single theme. This realization was the direct result of my then recent study of deception planning outside the military arena.[149] I had been looking at how target audiences' perceptions are manipulated by such seemingly diverse deceivers as counter-espionage experts, confidence tricksters, gambling sharps, pseudo-psychics, magicians, mystery-story writers, comedy writers, and practical jokers.

This research turned up an unexpected and interesting fact: Among all types of deceivers, only confidence tricksters, counter-intelligencers, and magicians possess and routinely use standard operating procedures for minimizing the unfortunate effects of discovery. Moreover, magicians have since 1785 gradually evolved a primitive but teachable theory. All other deceivers including soldiers leave themselves only the option of hasty flight, if even that.

---

149 Later published as J. Barton Bowyer [pseudonym of Bart Whaley & J. Bowyer Bell], *Cheating* (New York: St. Martin's Press, 1982).

Let us first see how that expert deceiver, the con artist, handles the problem.

4.2. The "Blow-Off" Technique

Confidence tricksters and other "hustlers" have developed a standard technique to buy enough time for a safe escape just before the victim ("mark") sees the operation for the scam it is. This technique is called "cooling out the mark" or the "blow-off". It is built into the final stage of the operation and is itself a deception.

Cooling the mark out is usually only a matter of convincing the disappointed and/or irate sucker that his loss involves neither loss-of-face nor was the responsibility of his "friend", the hustler. If played carefully, the hustler may even appear so blameless that he can hit the mark again later.[150]

In its most melodramatic form the blow-off uses the "cacklebladder". After the still unwitting sucker has been fleeced, his "partners" fake a police raid or stage a violent argument over the spoils, ending with one apparently dead in a pool of gore. The gore is merely animal blood that the "corpse" had concealed in a small waterproof bag—originally a "cacklebladder", a chicken's bladder. The "surviving" partners hastily agree to split up with the promise of meeting later to divide the wealth. The meeting, of course, never occurs and the sucker, believing himself an accessory to murder, may ever afterwards thank his good luck in having gotten out with his life. The 1973 Paul Newman-Robert Redford film, *The Sting*, ends with a cacklebladder blow-off that sends the villainous Robert Shaw scurrying off in fear, leaving his half million dollars with the suddenly very wealthy "corpses" of Newman and Redford. The 1986 film, *F/X*, is based on a further interesting switch on this con.

Magicians share with con artists the concept of the "blow-off". In fact, they borrowed this bit of jargon from the confidence tricksters. The blow-off in magic is simply a way of ending a trick with a twist that destroys any hypotheses that the audience may have developed as to the trick's method.[151]

Col. Dudley Clarke was the first military deception planner to reinvent the blow-off, which he called the "escape route". He recognized this problem and solution while dealing with double agents in the Middle East during WW2.[152]

Variations of the "blow-off" appear routinely in failed counterintelligence and counter-espionage operations. Their purpose is to make the opponent

---

150 Erving Goffman, "On Cooling the Mark Out," *Psychiatry*, Vol.15 (1952), 451-463; and Robert C. Prus and C. R. D. Sharper, *Road Hustler: The Career Contingencies of Professional Card and Dice Hustlers* (Lexington, MA: Lexington Books, 1977), 2.

151 Bart Whaley, *The Encyclopedic Dictionary of Magic, 1584-1988* (Oakland, CA: Jeff Busby Magic, Inc, 1989, entry under "blow-off".

152 Mure (1980), 42, etc.

uncertain that his own "penetrations" had been as successful for as long as he had thought. The "blow-off" strikes me as being well-suited to several kinds of para-military, guerrilla, and terrorist operations where failures can often be disguised by laying them at the door of another faction or group.

However, the "blow-off" has only limited application to conventional military operations. At best it would only leave the victim uncertain that a deception had even been attempted. And only then if the failure were effectively disguised as to its true purpose—a difficult proposition probably seldom worth the effort, particularly on a battlefield.

Let's now look at an SOP with unlimited application to military deception failure.

4.3. The Theory of "Outs"

Magicians have gone the con artist one better. In addition to the "blow-off" they have another highly developed concept, which in their extensive technical jargon, is called the "out". It applies to all magical deceptions. It could be systematically applied to all military deception operations; but, so far, it has not.

French amateur magician Henri Decremps foreshadowed The Theory of Outs in 1785 as Rule 2 in his list of principles of magic, where he wrote:[153] "Whenever possible have several ways of doing the same trick." This rule first appeared in English translation in an anonymous book, plagiarized also as Rule 2 but elaborated to include its reason:[154]

> "Endeavour, as much as possible, to acquire various methods of performing the same feat, in order that if you should be likely to fail in one, or have reason to believe that your operations are suspected, you may be prepared with another."

Magicians to this day observe Decremps' Second Rule, quoting or elaborating on it.[155]

The great French professional stage magician, Robert-Houdin, expanded on Decremps in 1868 when he wrote:[156]

---

153 Henri Decremps, *Testament de Jerome Sharp* (Paris: 1785).
154 *Parlour Magic* (London: Whitehead & Co., 1838), 123.
155 See, for example, John Henry Anderson, *The Fashionable Science of Parlour Magic* (6th ed., [London]: by the author, [c.1944]; *The Magician's Own Book* (New York: Dick & Fitzgerald), 79, taken verbatim from Anderson; Ellis Stanyon, *Conjuring with Cards* (London: L. Upcott Gill, 1898), 3.
156 Robert-Houdin, *The Secrets of Conjuring and Magic* (London: George Routledge and Sons, 1877), 32-33. This is the English translation of the original French edition of 1868.

> "However skilful the performer may be, and however complete his preparations for a given trick, it is still possible that some unforeseen accident may cause a failure. The only way to get out of such a difficulty is to finish the trick in some other manner."

He went on to explain:

> "However awkward the position in which you may be placed by a breakdown, never for one moment dream of admitting yourself beaten; on the contrary, make up for the failure by coolness, animation, and 'dash'. Invent expedients, display redoubled dexterity, and the spectators, misled by your self-possession, will probably imagine that the trick was intended to end as it has done."

Well, this sounds like Robert-Houdin was advising desperate improvisation in desperate situations. That may have worked OK for such a superb showman but surely not for average mortals. In fact he was a meticulous planner of his benign deceptions, leaving little to chance. Robert-Houdin should have added that the invention of expedients can and should be done in advance, anticipating the more likely types of failure. This oversight was finally filled in 1940 by the first manual of "outs", a perennial 79-page bestseller written by a wealthy Philadelphia amateur and aimed specifically at card workers.[157] A second, more generally applicable, 27-page pamphlet appeared three decades later.[158] While designed as wholly practical guides, both of these texts come close enough to stating a general theory that they can be applied to other fields of deception.

The most consistently successful modern conjurers design and practice an "out" for each step during a trick. Some go so far as to plan "multiple outs". And one of Dai Vernon's masterpieces is planned as a decision-tree each branch of which comprises optional outs.

---

157 Charles H. Hopkins, "Outs": *Precautions and Challenges* (Philadelphia: Chas. H. Hopkins & Co., 1940).

158 Steve Beam and Don Morris, *Inside Outs: Or For My Next Trick I'll Try One That Works* (U.S.: n.p., 1979).

5. Military Application of Outs

> When you put over a deception, always leave yourself an escape route.
>
> — Dudley Clarke as quoted in Mure (1980), 42

As the above quotation shows, Brigadier Dudley Clarke recognized the value of the "out." He is the only military deception planner I know of who rediscovered this. Moreover he taught this point to his staff.[159] However, he evidently taught them only by the inductive case-by-case method. He did not also explicitly state the underlying principle of how one goes about building this "escape route" into the plan.

The principle of the "out" can be systematically applied to all military deception plans to insure against their failure. The fact that it has not is simply a reflection of the primitive level of theory that prevails.

The "out" works in military deception for the same reason it works in any type of deception. This is the "Principle of Security of Options".[160] Deception, by its very nature, provides its own best security. Although the term "security of options" was original with me, the concept was merely a generalized extension of Liddell Hart's principle of "alternative objectives", first formulated by him in 1932.[161] Indeed, the magicians' Theory of Outs is simply a 1½ centuries earlier and more developed statement of this notion of alternative objectives.

Alternative objectives merely means that in designing an operation the Commander should have more than one goal or purpose in mind. Thus he may, like Rommel at the First Battle of Mersa el Brega,[162] keep open the option of defense or attack. Or he may be able to threaten attack at two points, his reserve positioned to exploit whichever one proves more vulnerable.

My notion of "security of options" extends Liddell Hart's principle of "alternative objectives"—that is, the geographical PLACE(s) targeted—to the other eight things that a deceiver can manipulate, namely TIME, STRENGTH, INTENTION, PAYOFF, PATTERN, STYLE, PLAYERS, and CHANNEL.

---

159 Mure (1980), 221.
160 First stated in Whaley, *Stratagem* (1969), 225-226.
161 B.H. Liddell Hart, *The British Way in Warfare* (London: Faber & Faber, 1932), 302, is the first appearance in print of his important theory & term of "alternative goals" or "alternative objectives".
162 Whaley, *Stratagem* (1969/2007), Case A26 (Mersa el Brega); and Whaley, "The Initiative as a Factor in Battle" (draft manuscript, 9 August 1987), 3-5.

For instance, the deceiver may keep the option of now-or-later timing. Indeed, for each of the nine types of things that can be manipulated, there are always, in principle, options or alternatives.

If the deception operation succeeds in anticipating the preconceptions of the target and playing upon them, reinforcing them, then deception security is absolute. In that case the victim becomes the unwitting agent of his own surprise, and usually no amount of warning (such as security leaks) can reverse his fatally false expectations. Even if the deception plan runs counter to or neglects playing upon the victims preconceptions, the very fact that it threatens alternative objectives will usually assure enough uncertainty to delay or defuse or otherwise blunt the victims response (as seen in Case 2.1.2, the Battle of Massicault).

The worst possible case would occur if the deception plan itself were prematurely disclosed to the victim. Such potentially disastrous disclosure is rare in general and unknown so far in battle. The closest to this was during the Battle of Midway in 1942 when faulty security permitted the U.S. Navy to see around the primitive Japanese deception operation and set an ambush.[163] A later example from strategic international politics was the disclosure in 1986 of the CIA's disinformation campaign against Libyan leader Khadaffi. At less exalted levels we see this in "tip-offs" of con games and other scams and commonly in "poison-pen" letters that reveal marital infidelities.

But, even if there were complete premature disclosure, all would not necessarily be lost. First, the disclosure itself would have to be believed. Second, if the deceiver knows or even suspects disclosure, he can actually capitalize on this by switching to one of the alternative courses of action or simply adopting a new deception plan to reverse appearances, as we saw General Alexander do in 1944 by successfully reversing an already successful deception operation by deliberately disclosing it. Even if the direction or objective of the attack has been compromised, the planner can still manipulate the victims perception of the timing or strength or even the intent or style of the attack.

---

163 See Whaley, *Stratagem* (1969/2007), Case A33 (Midway).

6. Priorities for Outs

In considering possible outs, the planner has two ways to go. He may look at the immediate situation and try to imagine all the possible options open to his commander and then compare these with his best estimate of enemy responses, juggling them about until one or more look like a good tradeoff. Indeed, I recommend this process as the most efficient one for generating the real plan, the real cover-plan, and the potential outs, all at one time.

Or the planner may chose a less intuitive and more systematic approach. Consider for a moment the structure of deception developed by Bell and Whaley and published elsewhere:[164]

The Structure of Deception

DISSIMULATION	SIMULATION
Masking	Mimicking
Repackaging	Inventing
Dazzling	Decoying

It is sufficient to say here that, in this model, the two main modes of deception and the three sub-categories under each are operationally defined in terms of how the "signatures" (identifying characteristics) of each real thing hidden and each false thing shown are differently manipulated.

It has been demonstrated that in practice as in theory, all three ways of hiding the real (dissimulation) can accompany the three ways of showing the false (simulation) in any of their nine possible combinations. Here the three ways to dissimulate and the three to simulate have been arrayed in descending order of assumed effectiveness. The hypothesis of my collaborator, Dr. J. Bowyer Bell, was that the most effective way to dissimulate would be by *masking*. Therefore, if masking fails to confer enough invisibility to the real object or event and it is detected, the deceiver can then resort to *repackaging* to disguise it. And if repackaging fails and the thing is recognized for what it is, then *dazzling* can be used as a last-ditch measure to at least confuse the target about some characteristics. Similarly, Bell assumed that the most effective way to simulate would be by *mimicking*. Therefore, should mimicking fail to show the false by a convincing imitation, the deceiver can resort to *inventing* to create an alternative reality, And if inventing also fails, *decoying* can still be used at least

---
164 Whaley (1982), 186.

to attract the otherwise undivided attention of the target away from it. Bell suggested that deception planners not only also have the possibility of an out but indeed, a series of priorities for them—the poor, better, best in each of the two columns of the above chart.[165]

Consequently, I thought it likely that the most effective deceptions will dissimulate by masking and simultaneously simulate by mimicking, while the least effective deceptions would be those that combine dazzling with decoying to achieve only a mere razzle-dazzle effect. Therefore, while dazzling-decoying deceptions might get invented, few if any would survive frequent operational experience and be soon dropped from the repertoire.[166]

To test Bell's hypothesis and my subsidiary one, I chose to use the tricks of magicians, because among all professional deceivers, they have by far the most frequent experience of both failed and successful deceptions. Accordingly, 60 magic tricks that had been collected to illustrate a previous paper[167] were fitted into a 3x3 matrix as follows:[168]

The Matrix of Deception
(as applied to 60 conjuring tricks)

DISSIMULATING	SIMULATING		
	Mimicking	Inventing	Decoying
Masking	19	10	10
Repackaging	4	3	1
Dazzling	10	2	1

As can be seen in practice or at least in magicians' practice, both hypotheses are substantially verified. Masking and mimicking are not only overwhelmingly the most common methods used for dissimulation and simulation respectively; but they also are the two used most often in combination. Conversely, only one example of dazzling-decoying was discovered (the Ultima Thule card location), despite an even wider search among magic tricks. Thus, this is an apparently useful procedural guide for designing the potentially most effective types of deception and avoiding the lesser and least effective ones.

Ruses of this dazzle-decoy type are also quite rare outside of conjuring. For example, of the dozens of carnival "flat store" (cheating) games of pretended

---

165 Whaley (1982), 187.
166 Whaley (1982), 187.
167 A draft paper by myself written in 1981 that used an "opportunity sample" of 60 magic tricks. This should be replicated using the much larger number of tricks now collected. Then a parallel analysis of military cases could be done and the two studies compared.
168 Whaley (1982), 187-188.

chance or skill, only one is of this type. Carnies call it, appropriately, Razzle-Dazzle (but only among themselves), the suckers knowing it by such disguised names as Bolero, Double Up, or Ten Points. This "game" uses a method of scoring too complicated for the sucker to follow in the short time permitted so that the "flat-joint" operator need only false-count to make the sucker lose.[169]

A classic military example of dazzle-decoy is the "Two Pattens Ruse" described elsewhere.[170] Another was the "dazzle painting" on warships and freighters in World Wars I and II.[171] To repeat, this combination is the least sure and least effective type of deception and should only be tried as a last recourse. Col. Ralph Ingersoll understood this, even to the point of denying that his razzle-dazzle Two Pattens Ruse was true deception.[172] The best deceptions make the victim certain but wrong. Dazzle-decoy merely confuses him; and a confused, even panicky opponent just might act contrary to your wishes. But even so, much better to instill confusion than nothing.

This model of hiding and showing can also be turned around, as I have explained in another manuscript, to serve as a guide for counter-deception analysts in assigning priorities in their search for deception.[173]

---

169 John Scarne, *Scarne's Complete Guide to Gambling* (New York: Simon and Schuster, 1961), 478-489.
170 Whaley, "Practice to Deceive", draft RAND paper, 1987.
171 Seymour Reit, *Masquerade* (New York: Hawthorne Books, 1978), 194-197.
172 Author's interview with Ingersoll, 12 May 1973.
173 Whaley, *The Maverick Detective: The Whole Art of Detection* (manuscript in progress since 1986).

7. Beyond the Out: Asymmetries between Deceiver & Target

> "We carry on acting dumb and innocent. If they think we're dumb and innocent, they'll get careless."
>
> — Lee Child, *Killing Floor* (NY: Putnam's, 1997), Chapter 16

During the final editing of this paper, I was discussing the problem of deception failure with American professional magician Ben Robinson. We talked of the conjuror's need to recover from the occasional real failure of either manipulation or apparatus. As the author of a recent biography of a CIA contract magician in the early Cold War years,[174] Robinson was fully aware that the concept and practice of the Out is equally relevant to the planning of political, intelligence, military, and magical deceptions.[175] But he soon surprised me by taking our conversation onto an oblique path. Robinson's repertoire includes a comedy magic act. In describing it, he explained that in addition to the usual conjuror's Outs, he adopted the stage persona of an *incompetent* comedy magician. This lulls his audience by leading them to expect failure. But this, of course, is just a way of hiding his expertness. He appears to fumble and fail, then succeeds in his trick. This set me thinking—with the following result:

The Out is a tightly focused *tactic* adopted to recover as best as possible from one's failed deception. It is pre-planned but only put into play after failure. In contrast, the Role a deceiver plays when attempting to deceive the target is a broadly conceived *strategy* adopted beforehand to enhance the chances of success.

Rule:

Whenever a competent professional deceiver—magician, con artist, soldier, etc—confronts a particular audience or target—it is essential that the deceiver have at least some sense of that audience's level of

---

174 Ben Robinson, *The MagiClAn: John Mulholland's Secret Life* (USA: Lybrary.com, 2008). On Mulholland see also H. Keith Melton and Robert Wallace, *The Official CIA Manual of Trickery and Deception* (New York: Morrow, 2009).
175 On the morning of 24 Jun 2010 at the Magic Castle Hotel, Hollywood.

competence for detecting deceptions. In that case the wise trickster will chose whichever level of *apparent* competence or incompetence is most appropriate for "working" that audience. If, uncertain of your target's detection ability, test for their level of gullibility. That was the learning curve of the British LCS and Double-Cross Committee deceivers in WW2. And all "kid show" magicians soon learn that one reason children make tough audiences than adults is that, having not yet become fixed in their understanding of how things work, they are less likely to accept the false premise or hypothesis that the magician is trying to sell.[176]

I recognize three main levels, styles, or roles that define deceivers:

1) **The Inept.** Implies a near 100% failure rate, zero inability to deceive, no expectation of success. As in magician Ben Robinson's case, this is the incompetent bungler. Not just the usual comedy magician who accompanies "straight" magic with comedic patter, but that sub-type who deliberately plays the Fool.

    - This is the court jester who, by playing Fool, can sometimes trump the king's supposedly wise counselors.[177]
    - Magicians will recognize Cardini's drunk act or Carl Ballantine's hopelessly disorganized act.
    - Intelligencers will recognize Graham Green's Mr. Wormold, that bumbling vacuum cleaner salesman who, coopted as a British secret agent, finally outwits Cuba's Intelligence Chief.
    - Con artists and other charlatans play either Honesty Personified or Honor Among Us Thieves roles. In either case, the role Implies zero deception. If a deal goes sour, well, that's just bad luck or somebody else's fault. We all know Ponzi and Madoff. Intelligencers will recognize the Penetration Agent or Mole (but not the Double Agent). One sub-category used by some confidence artists to throw off any suspicion of deceitfulness is to play the Rube or Hick—pretending to be naive or dull-witted.[178] Again this role sometimes gives a built-in Out (although a rather weak one).

---

176 Henry Hay, *The Amateur Magician's Handbook* (New York: Crowell, 1950), 3-4.
177 Beatrice K. Otto, *Fools are Everywhere: The Court Jester Around the World* (Chicago University Press, 2001).
178 The classic example is the "rube act" of William "Canada Bill" Jones, the most famous mid-19th century Three-Card Monte player, as described by his sometime partner in the con, George H. Devol, *Forty Years a Gambler on the Mississippi* (Cincinnati: Devol & Haines, 1887), 191-216, 220-221, 285-287.

2) **The Average.** Promises a mediocre success rate—somewhere around a lackluster 20%-80%.

- The prime example is the Mentalist, also called a Mind Reader or Mental Magicia, who promises at most an 80% rate. This is the world of Kreskin or even Uri Geller. By admitting beforehand that they're sometimes wrong, they give themselves a built-in Out. But, because they are only pretending weakness, they win unearned credit when they do get it right. Consequently, this is a very powerful technique.
- Among detectives—police and private—the great majority are merely savvy hard workers like Hammett's Sam Spade or Gores' Dan Kearny & Associates.
- Among intelligencers, this is their usual situation, one mirrored in the semi-fictional worlds of Le Carre's George Smiley or Maugham's Secret Agent. Not those of 007.

3) **The Master.** Promises a 0% failure rate. To sustain this claim, the Out is essential.

- Among Master Magicians we have Houdini or David Copperfield.
- Among police detectives the Canadian Mounties and American FBI "Always Get Their Man!" And, among the fictional super detectives, the British have Doyle's Sherlock Holmes, the French have Gaboriau's Monsieur Lecoq, and even the Republican Chinese had Ch'eng Hsiao-ch'ing's scientific detective, Huo Sang; and even the Chinese Communists have their own detective hero.
- Among major intelligence services near superhuman excellence was a legend encouraged by the British Security Service (MI5) and Secret Intelligence Service (MI6), the Russian KGB, the CIA, and the Israeli Mossad. The British intelligence services enjoyed an enormous reputation since Queen Elizabeth I's Sir Francis Walsingham, one that persisted into times of great weakness as in 1914 at the outset of World War One when MI5 mustered only 17 souls, including Director Kell and one caretaker.[179] Similarly, as I reported in 1969, the various Soviet intelligence services were, while indeed the world's largest, most intrusive, and most far-flung, were far less competent than generally portrayed by

---

179 As reported by MI5's official historian, Andrew (2009), 29. On the gulf that separated the realities of the British intelligence services from reality see also Robert Dover and Michael S. Goodman (editors), *Spinning Intelligence: Why Intelligence Needs the Media, Why the Media Needs Intelligence* (London: Hurst, 2009), 185-219.

non-communist writers.[180] At its extreme, this is the ridiculous & bombastic wannabe fantasy perpetuated internationally by James Bond and in the USA by Jack Bauer.

Each of these three roles simulates one particular level or style, one that the deceiver is wise to tailor to fit the target's known or presumed level of detection competence. As the target's skill level will also fall somewhere along the Inept-Mediocre-Master continuum, the possible interactions between deceiver and target can be roughly mapped onto a 3x3 matrix. Of the nine cells in this matrix, only three give nearly symmetrical matches between the skill levels of the two players. The other six cells are clearly asymmetrical.

Obviously, this requires that the deceiver always make a strong effort to discover the opponent's skill level at deception analysis. And, of course, all deception analysts should do the same. Obviously, a great advantage is enjoyed by whichever party has the better knowledge of its opponent's skill level—and their own!

This point was explained by Canadian-American master magician Jeff Busby:[181]

> **"I never perform tricks for other magicians unless I have discovered their skill levels and styles of thinking."**

Busby's rule is worth stressing. It puts into words two important facts that all top conjurors (and con artists) know at least intuitively. First, that when faced with a lay audience, magic is a game they play without any clear notion of the rules. Then the magician wins every time. However, whenever one magician "works" an audience of other magicians, Busby's rule must be closely applied. And it must be applied.

Thus the most highly and consistently effective deceivers will go beyond the Out by adopting an asymmetric strategy. Conversely, the most effective deception analysts must be sensitive to this possibility when assessing any situation where such tricks might be in play. But who has the advantage? Who is the deceiver, who the dupe? We or them? There are two basic and opposing answers. The first is unnecessarily complex—a veritable tangled web, an infinite regression into that wilderness of mirrors. I'll leave its solution to game theorists and wish them more than their usual ability in developing practical guidelines for real-world situations.

The second answer is very simple. Simulate a mix of weakness and strength. This applies whenever you are unsure of your skill level at either deceiving

---

180 Whaley, *Soviet Clandestine Communication Nets: Notes for a History of the Structures of the Intelligence Services of the USSR* (Cambridge, Mass.: Center for International Studies, MIT, 1969); and Whaley, *Codeword BARBAROSSA* (Cambridge, MA: The MIT Press, 1969). 2.
181 Jeff Busby, via telephone to Whaley, 22 Jun 1999.

or detecting and/or are unsure of your opponent's skills. Then, adopt a mixed strategy of pretending weakness *and* strength. But this strategy has two interacting parts. First, pretend that your organization or unit is a blend of several inept, many mediocre, and a few master deceivers (or deception analysts or both). Sound familiar? Naturally, for it accurately describes every intelligence service I've studied. Second, pretend that even your few Masters have their "off" days or, better, are not super-competent in all aspects of this job. Again, this is realistic.

Although there may be positions, as with the con artist, where one might want to seem much *less* competent than you are, the optimal position will most likely be to pretend to much *more* competence than you have. Why? Because by greatly *underplaying* your skills you may lull your opponent into a trap. That's good *offensive* deception. But, to the extent that you can greatly *exaggerate* your skills, you render your opponent uncertain and, therefore, more cautious, less active, and slower. That's good *defensive* deception. Either way, this is a happy choice.

This mixed strategy assures a kind of Out whenever one of your deception operations has failed *and the opponent knows it*. They will then most likely attribute this failure to some procedural flaw in your intelligence system, one or two blundering individuals, shoddy teamwork, or just a bad day on the part of any Master that might have been involved. Even then, the worst outcome is that you and your opponent will, temporarily, be equally proficient at deception or its detection—a "wash" with no harm done. But this is no excuse to abandon deception and deception analysis. To abandon those skills entirely only assures that you will become every deceiver's perpetually gullible dupe.

There is, unsurprisingly, a more realistically detailed third answer. It takes the simple or default answer as its start point and evolves theory in lock-step with hard digging for data and tight analysis. But those are tasks beyond this paper's remit.

This chapter has explored into largely uncharted country and only sketched its outlines. It would be useful to have a monographic study that develops and blends two key elements. First, a refinement of our existing metrics for measuring the degrees of effectiveness of deception and its detection.[182] Second, an elaboration of this chapter's model of the levels & asymmetries of deception and Its detection.[183]

---

[182] As initially developed in Whaley, *Textbook of Political-Military Counterdeception: Basic Principles and Methods* (FDDC, 2007), 72; Whaley, Stratagem (1969/2007), Chapter 5.
[183] As analyzed in Whaley, *Textbook* (FDDC, 2007), 40-42.

Bibliography

This bibliography gives only those few books and papers that have been identified as explicitly addressing the question of deception failure. All other citations are traceable through the footnotes. The code "LOC: BW" means that the book or article is in the Whaley Collection in the Barton Whaley Deception Research Center, a room in the CIA Library, Langley, Virginia.

Andrew, Christopher (1941-)
> *Defend the Realm: The Authorized History of MI5*. New York: Knopf, 2009, xxiii+1032pp.
>
> LOC: BW.

Balfour, Michael (1908-1995)
> *Propaganda in War 1939-1945*. London: Routledge & Kegan Paul, 1979, xvii+520pp.
>
>> I'll let this one book represent the many detailed & well-documented studies of psychological operations in war. This topic is closely related, even often allied to military and strategic deception. However, it has proved to be sufficiently less consequential to the outcome of strategic political and military operations than its full-bore military counterpart. Consequently, I give it only occasional passing mentions here. A more comprehensive study would benefit from a study of the many psyops aborts, failures, and blowbacks.

Beam, Steve (1958-); Don Morris
> *Inside Outs: Or For My Next Trick, I'll Try One That Works*. No place: no publisher, 27pp.
>
> LOCATION: BW.
>
>> A pamphlet on the magician's "Out". A supplement to Hopkins (1940).
>>
>> Steven L. Beam is a well-known American semi-pro magician. Morris was a tennis pro and amateur comedy magician.

Cohen, Eliot (1956-); and John Gooch (1945-)
> ★ ★ ★ *Military Misfortunes: The Anatomy of Failure in War*. New York: The Free Press, 1990, viii+296pp. LOC: BW.
>
>> Although a fresh, commendably researched, and often convincing effort to account for military failure, it is one that stresses doctrinal

and organizational mismatches with the enemy. However, its tilt toward the Wohlstetter-Betts's "no-fault" notion of the inevitability of surprise (pp.40-41, 117-120, 181) largely excludes the whole factor of deception.

Cruickshank, Charles (1914-1989)
Deception in World War II. Oxford: Oxford University Press, 1979, 248pp.

LOC: BW.

One of the earliest and consequently influential accounts of the military deceptions of WW2 to appear following the first wave of bulk declassification of British documents on that topic which began in 1978. Misleadingly titled—it's limited to *British* deception operations and then only in the West Europe and Mediterranean theatres, with merely passing recognition of the American and German deception operations. Moreover, it doesn't even bother to take into account the German & Italian perceptions of Allied deception operations, despite the then public availability of much of their documents. Although based largely on official British records, the author under-plays the value of deception operations, particularly any that fell short of their intended goals.

Deception Research Program, Everest Consulting Associates, Mathtech, Inc, Office of Research and Development, ORD)/CIA, Analytical Methodology Research Division
Deception Failures, Non-Failures and Why. Washington, DC: Office of Research and Development, CIA, Jan 1982, ii+72pp. LOC: BW (copy); CIA Library (HIC.BF.637.D4.U56.1982); scarce.

Provides 11 specific case studies from 1939 through 1944 where deception failed. However, these illustrate only a few of the causes of the failures to which the author is otherwise aware of. Nevertheless, this study provides the earliest detailed analysis of the causes of both failure and success of deception operations. Partly copied (without credit) by *FM 90-2 ("Battlefied Deception")* (1988), which by omitting the 3 psyops deception cases in Maxim reprints only 8 "military" case studies.

I believe this anonymous paper was largely or entirely written by L. Daniel Maxim of Mathtech/Everest. And that is his best recollection as confirmed in a conversation with BW on 27 Apr 2010. Mary Walsh, a mathematical topologist, was the then head of the Analytical Methodology Research Division.

Ferris, John (1957-)
: "The Intelligence-Deception Complex: An Anatomy," *Intelligence and National Security*, Vol.4, No.4 (Oct 1989), 719-734.

LOC: BW (copy).

> Ferris attempts to adjudicate the relative merits between Klaus-Jürgen Müller (1987) and Michael Handel (1987a) on the problem of how to assess the success or failure of a deception operation. The author does this from the viewpoint of the interface between historiography and epistemology. In Ferris's judgment (and mine) Handel wins by a length. However and unfortunately, like both Müller and Handel, Ferris's otherwise useful analysis founders because he fails to recognize that surprise is almost never an either/or phenomenon.

FM 90-2 (Battlefield Deception)
: Washington, DC: Headquarters, Department of Army, 3 Oct 1988, various pagings [= ca.168pp].

LOC: BW (copy); Internet.

> An official US Army manual, better than most but showing only a weak understanding of deception principles or practice. The section in Chapter 7 on deception failures was seemingly copied (without credit) from Maxim (1982).

Drafting began in late 1987 and reached its final approved draft in Dec 1987. Superceded the edition of 2 Aug 1978.

Gerwehr, Scott (1968-2008); Russell W. Glenn (1953-)
: *The Art of Darkness: Deception and Urban Operations*. Santa Monica: RAND, [2000], xviii+70pp. LOC: BW.

> Includes (pp.34-36) a brief but useful section titled "What Are the Dangers of Employing Deception?" See also their *Unweaving the Web* (RAND, 2002).

Goffman, Erving (1922-1982)
: "On Cooling the Mark Out," *Psychiatry: Journal of Interpersonal Relations*, Vol.15, No.4 (1952), 451-463. LOC: Internet.

> On the con artist's version of the magician's "out". Dr. Goffman was a prominent Canadian-American sociologist.

Gopnik, Adam (1956-)
"The Real Work: Modern Magic and the Meaning of Life," *The New Yorker*, Vol.84, No.5 (17 Mar 2008), 56-69. LOC: Internet.

> A humorous but perceptive introduction to conjurors as mind-game players. Explains such key points of magical theory as "The Real Work" and Rick Johnsson's "The Too Perfect Principle". See also Johnsson (1970).

> Gopnik, an American writer & essayist has contributed to *The New Yorker* since 1986.

Handel, Michael (1942-2001)
"Strategic and Operational Deception in Historical Perspective," in *Handel, Strategic and Operational Deception in the Second World War* (London: Frank Cass, 1987), 1-91.

> Dr. Handel had been Professor of National Security Affairs at the U.S. Army War College from 1983 to 1990 when he moved to his final posting as Professor of Naval Strategy at the U.S. Naval War College.

Hanna, Christopher H. (1962-)
"From Gregory to Enron: the Too Perfect Theory and Tax Law," *Virginia Tax Review*, Vol.24, No.4 (Spring 2005), 737-796. LOC: BW; Internet (ProQuest).

> Surprisingly, the best account of magicians' Too Perfect Theory, although only intended to apply it to tax law. Takes its main examples in the realm of high finance fiddling those landmark cases of Gregory vs Helvering in 1935 and the Enron Corporation in 2001. See also Johnsson (1970).

> Dr. Hanna (1988 U of Florida JD) has been a Professor of Law at Southern Methodist University since 1990. A most knowledgeable amateur magician, he supplemented his own deep research by discussions with such senior conjurors as Jamy Ian Swiss & Richard Hatch.

Hart-Davis, Duff (1936-)
Peter Fleming: A Biography. London: Jonathan Cape 1974, 419pp.
LOC: BW.

Holt, Thaddeus (1929-)
The Deceivers: Allied Military Deception in the Second World War. New York: Scribner, 2004, xx+1148pp.

LOC: BW.

Hopkins, Charles H. (1899-1948)
"Outs": Precautions and Challenges. Philadelphia: Chas. H. Hopkins & Co., 1940, 79pp.

LOC: BW.

> The classic pamphlet on the subject of the magicians' "out". Although designed for conjurors specialized in card work, it is relevant to all branches of magic and, indeed, all forms of deception. Highly recommended by master magicians Jeff Busby & Abb Dickson. Supplemented by Beam & Morris (1979).

> Ghosted by Walter S. Fogg (1890-1942). Hopkins, the nominal author, was a Philadelphia philanthropist and amateur magician who paid for publication of this monograph. Howard, Michael (1922-)

★★★ *British Intelligence in the Second World War*, Volume Five ("Strategic Deception"). New York: Cambridge University Press, 1990, xiii+271pp.

LOC: BW.

Johnsson, Rick (1937-1989)
"The 'Too Perfect' Theory," *Hierophant*, No.5/6 (Fall/Spring 1970/71), 247-250.

LOC: BW (copy).

> Frederick "Rick" Johnsson was an American teacher and amateur close-up magician, best known as a columnist for *The Linking Ring* magazine from 1976 until his early death in 1989. See also Hanna (2005) and Gopnik (2008).

Lewin, Ronald (1914-1984)
The Chief: Field Marshal Lord Wavell. London: Hutchinson, 1980, 282pp.

Maxim, L[eslie] Daniel (1941-)
SEE UNDER: Deception Research Program (1982).

Müller, Klaus-Jürgen (1930-)
"A German Perspective on Allied Deception Operations in the Second World War," in Michael Handel (*editor*), *Strategic and Operational Deception in the Second World War* (London: Frank Cass, 1987), 301-326.

LOC: BW (in MH).

A revisionist (pessimistic) view of the Allied deception operations that is largely disproved in editor Handel's introductory essay. Müller was a leading German military historian.

Mure, David (1912-1986)
Practise to Deceive. London: William Kimber, 1977, 270pp.

LOC: BW.

A memoir-history of "A" Force, the British military deception planning team in Cairo during WW2.

Master of Deception: Tangled Webs in London and the Middle East. London: William Kimber, 1980, 284pp.

LOC: BW.

Biography of Brigadier Dudley Clarke (1899-1974) who, as commander of "A" Force in Cairo, was chief of British military deception planning in North Africa and the Middle East during WW2.

Vego, Milan N. (1941-)
Joint Operational Warfare. Newport, RI: Naval War College, 2007, various pagins totaling 1492pp.

LOC: Internet.

Includes a brief section on "Reasons for Failure" on pp.VII-43 to VII-115. Dr. Vego (1981 George Washington U PhD in History) was born & raised in Yugoslavia. The USA granted him political asylum in 1976 and citizenship in 1984. He has been Professor of Operations at the Naval War College since 1991.

Whaley, Barton (1928-)
Stratagem: Deception and Surprise in War. Cambridge, MA: Center for International Studies, MIT, 1969, ix+263+628+49+ [25] leaves [=985pp].

LOC: BW.

Reprinted (less Appendix B), Boston: Artech House, 2007, xxvii+560pp.

LOC: BW.

> A comprehensive study. Following an historical survey of deception in the worlds major nations, systematically analyzes 168 battle in 16 wars from 1914 thru 1968.

"Toward a General Theory of Deception," *Journal of Strategic Studies* (London), Vol.5, No.1 (March 1982), 178-192.

LOC: BW (offprint).

Whaley, Bart; and Jeffrey Busby (1954-)
"Detecting Deception: Practice, Practitioners, and Theory," reprinted in Roy Godson & James J. Wirtz (*editors*), *Strategic Denial and Deception* (New Brunswick: Transaction Publishers, 2002), 181-221.

LOC: BW (G&W).

> The first general theory of detection, originally published in 1999. Busby contributed the key technique (described on pp.212-218) that I call "Busby's Ombudsman" and later characterized as an "intuitive trigger".

Appendix: Extracts from 1982 CIA contract paper titled *Deception Failures, Non-Failures and Why*

Everest Consulting Associates, Mathtech, ORD/CIA (Analytic Methodology Research Division), *Deception Failures, Non-Failures and Why* (Washington, DC: Office of Research and Development, Central Intelligence Agency, Jan 1982, ii+72pp). I believe the principal (or only) author was Dr. L. Daniel Maxim of Everest/Mathtech. The CIA sponsor was Mary Walsh.

I. Introduction

[*The 5-page Introduction is omitted here, as substantially superceded by more nearly complete access to data from both the deceivers & the deceived.*]

II. Examples

[*11 examples of failed deception were given in 38 pages. Together with their corresponding case numbers in Whaley's above text these are:*

Operation HIMMLER	Case 1.3.2.1
Battle of Britain	p.5n2
SEA LION	Case 1.3.3.1
ALBION	Case 2.6.1
Black Propaganda	Case 2.3.10
ELEPHANTIASIS	Case 1.3.2.2
Soviet Tactical Radio Deception	Case 1.4.2.2
COCKADE	Case 2.2.6
ACCULATOR	Case 1.4.2.3
IRONSIDE	Case 1.3.3.2
Anzio	Case 2.2.9]

III. Failure vs. Success

[*The original 6 pages are reproduced word-for-word here.*]

When Deception Fails: The Theory of Outs

In spite of a comprehensive search of military history, it has been difficult to find examples of clear deception failure. There are a number of possible reasons for this. First, since no one, particularly a military commander or bureaucrat, likes to admit failure, the perpetrator is not likely to volunteer the information or evidence of his shortcomings. Nor is the intended victim. Although he may be tempted to demonstrate how clever he was in seeing through the ruse, it may have been done with the aid of some sensitive intelligence success which he prudently realizes might be needed in the future. Alternatively, the deception may have been so poorly presented to the intended victim that he missed it entirely. Finally, record keeping of a deception operation often degenerates as it progresses. In the event of a failure, the available documentation may be inadequate to reconstruct the reason or details or the evidence intentionally destroyed to preclude embarrassment. Even considering these essentially administrative and psychological reasons however, the real reason is that, in comparison to the attempts, there really aren't very many. The classified list is not much longer and scarcely more exciting. It can accurately be stated that deception nearly always succeeds, at least to some degree. Indeed it should be emphasized that deception may succeed even when one or more of the previously listed causes for failure is present.

For example, the German agent CICERO, as the valet of[113] the British ambassador to Turkey, photographed sensitive documents borrowed from the ambassador's bedroom safe and passed them to a German military attache. Controversy remains over whether CICERO was initially planted on the Germans for deception purposes or was simply caught and turned through an intelligence success via TRIANGLE.[114] It is probable that he was a plant from the beginning but, in any case he was indeed controlled and at least some elements of the Abwehr became suspicious when one set of negatives contained images of fingers of *both* hands, a tricky feat in supposedly unassisted clandestine photography. A German photographic expert evaluated the film and confirmed the[115] probability of deception only to be ignored. Although von Ribbentrop apparently suspected CICERO's material also, a large number of individuals having access including Hitler apparently considered it genuine. **[BW: It is now fairly certain that CICERO was never under British control.]**

Such bureaucratic weaknesses have existed and continue to exist in many if not all intelligence services. At the lowest level the phenomenon is termed "the love affair between a case officer and his agent." Recruiting and running an agent takes such skill and emotional involvement that the case officer's personal and professional stature become closely involved with and dependent on the agent's continued performance and, implicitly, his credibility. The case officer then, is strongly motivated to maintain that credibility and tends, at least subconsciously and even actively in some cases, to ignore evidence of control.

It may even reach the point that the case officer is concerned for his personal safety if deception and control of his agent should be detected. During World War II the Germans at least threatened their case officers with "duty on the Eastern Front" for such a blunder and the Soviets are reportedly no less understanding. Western bureaucratic revenge is generally limited to slowed promotion or transfer to less desirable assignments or out of the counterintelligence field altogether. While this is seemingly less threatening the motivation to continue an otherwise promising career is understandably strong, particularly when viewed in the context that one piece of information more or less "won't really make that much difference in the overall scheme of things."

At the management levels of the intelligence structure the emphasis shifts to the information the agent provides. If it is unique enough the policymaker becomes aware of the nature of the source at least in general terms and begins formulating his policy on the basis of that information. Also, having been allowed a glimpse of the real world upon which spy fiction is based, many policymakers become understandably intrigued with the process of espionage. Counterintelligence managers sometimes trade on this, sharing details of operations with the policymaker to gain and maintain his interest and support. The policymaker, in turn, often tends to attach greater credibility to information[116] gathered by such a process apparently at least partly because of its sensational nature and is thrilled by the bureaucratic, political and personal power attached to such limited access.

The intelligence manager then has a vested interest in maintaining the flow of that information and the chain of self-deception is finally complete when the policymaker finds himself reluctant to admit that some of his policy decisions may have been based on flawed information. This is particularly true if that information has tended to support his strongly held convictions. Thus, when a single source of information supplies major or the only available intelligence on a particular topic, it forms a built-in vulnerability to deception in spite of obvious errors in its structure or execution.

It appears then that both man as an individual and man collectively organized into governments and sub-elements of governments such as intelligence services and policymaking bodies are highly vulnerable to deception and self-deception.

Organizations are composed of individuals who can be deceived on a one-to-one basis. However the organizational process itself suffers an inherent vulnerability to deception and certain functions tend to increase that weakness.

- Bureaucratic inertia—the resistance to change

- Layering—the insertion or existence of numerous levels of management or supervision between the analyst who should or does detect deception and the policymaker who should see through it and make decisions accordingly

- Self-protection of all echelons against outside influences

- Member actions—individuals tend to act and make decisions in a pattern placing self-interest first, task or goal next, both ahead of the overall organizational or national interest.

Intelligence analysts, charged with the detection of deception, also at times appear to demonstrate a vested interest in minimizing the importance of, or not finding at all, evidence to support the existence of deception. The primary reasons are:

- Professional pride—Many analysts consider it a personal affront to their ability to suggest that they could be fooled in the assessment of intelligence information. These same individuals often tend to play down the scope, importance and effectiveness of foreign deception.

- Reluctance to add uncertainty—The intelligence process is intended to provide policymakers and military decisionmakers with information upon which they can place high confidence. Any factor which adds uncertainty to the system is obviously unwelcome to the point where analysts are often tempted to avoid or downgrade the consideration of deception as a potential influence.

- Bureaucratic resistance—For similar reasons the bureaucracy avoids the consideration of any topic which might complicate management of the estimating process or the policymaker's acceptance of the product.

- The evidence is contrary to some institutional or personal model of reality.

It would seem, and history bears it out, that deception as defined in the classic sense nearly always succeeds at least to some degree. It is largely a victory for self-deception. The target has deluded himself into believing or disbelieving as the deceiver wishes. Either clever planning and execution by the deceiver or fortuitous (for the deceiver) events or a combination of both have conspired to bring about this misperception. Thus the deceiver only facilitates the victim's self-deception or misperception. If those misperceptions agree with or support other previously held evidence or convictions to the point where otherwise contradicting evidence is downgraded or ignored the deception is more likely to be successful in spite of technical flaws or errors in execution.

IV. Why Does Failure Occur?

[*The original text of this 1-page chapter is reproduced here, omitting only Figure 11.*]

Having laboriously identified such a large number of technical and bureaucratic reasons for deception to succeed why then does it fail, if ever so seldom? The answer apparently lies in the presence of the right combination of deceiver technical error, inadequacy and overall implausibility of the deception *and* target vulnerability to self-deception, graphically depicted below (figure 11). In general, the process shows greater sensitivity to changes in target vulnerability than in technical perfection.

Figure 11: Technical Quality of Deception vs. Target Vulnerability
[*see original text*]

V. What Can Be Done?

[*The original text of this 5-page chapter is reproduced in its entirety here.*]

Measures to minimize the impact of deception on an intelligence system fall into three categories which overlap to some degree:

Operational—Operational measures are generally employed in the intelligence tasking and collection phases to minimize the probability of collecting deceptive information without its being recognized or recognizable as such. They are normally but not exclusively procedural.

Technical—Technical measures are aimed at assisting the collection and analytical processes in identifying deceptive information. They are normally system or hardware oriented. Further discussion of these areas is impossible at the unclassified level. A third and broader area can be discussed, at least in theoretical terms.

Institutional—Institutional measures are those steps that can be taken by all or portions of the intelligence and policymaking apparatus to minimize vulnerability to and the impact of deception.

The first step must be to sensitize the entire spectrum of information collection, processing, analysis and consumption to the probability and nature of deception. Instruction courses designed for the particular level and audience would be a major step to bring about an understanding and acceptance of the phenomenon.

Second, it is essential to proof the intelligence collection and processing elements of the bureaucracy against penalties, real, implied or imagined, for errors or deficiencies in the process of detecting and analyzing foreign deception's impact on the intelligence, decision and policymaking processes.

Both steps require specialized organizations staffed with full time personnel experienced in the theoretical, historical and operational aspects of deception in the broadest context. These organizations should respond directly to the highest level of authority in their particular element of government. The individuals assigned to such programs should be volunteers, rewarded appropriately with promotion in spite of separation from their peers. Each must be given the most rigorous security screening and, once accepted, be given virtually unlimited access to intelligence regardless of its sensitivity. To do otherwise is to frustrate the analytical effort unnecessarily. Finally, not enough can be said for security in terms of the technical measures, techniques, successes and failures of the deception program. In virtually all cases, foreign knowledge of one's deception detection program can permit the adversary to redirect his efforts much more rapidly and cheaply than new countermeasures can be developed.

[*TABLE:*] **Reported Deception Failures**

[*This 2-page summary table is omitted here.*]

End Notes

[*Of the 11 pages of 127 end notes in the original work, only those from Chapter III are reproduced here.*]

113. Sir Hughe Knatchbull-Hughessen was a very senior diplomat. In the British tradition, at least at that time, he was kept informed of a somewhat wider scope of sensitive information and activities than that required for his assignment in Turkey. In addition to information regarding the Mediterranean he passed the codename for the genuine invasion of Europe, OVERLORD.

114. TRIANGLE was the codename for decrypted German wireless traffic dealing exclusively with intelligence matters. Encipherment was either by the ENIGMA machine or by special Abwehr manual ciphers. (DRP Glossary/Mure).

115. David Mure, *Practise to Deceive*, (London: William Kimber, 1977), p. 109.

116. The Soviets are apparently firmly wedded to this concept. John Barron (The KGB: The Secret World of Soviet Secret Agents, p. 191, 192) writes

"... KBG intelligence collection in the main does entail illegal methods. For the KGB continues to rely in the extreme upon the individual clandestine agent, the human spy, rather than upon technical means and deductive analysis."

"... [General Alexander] Orlov notes that the Russians consider only information so obtained as real *intelligence* and regard all other information as mere *research data*. 'According to the views of Russian officers, it takes a *man* to do the creative and highly dangerous work of underground intelligence on foreign soil; as to the digging up of research data in the safety of the home office or library, this can be left to women or young lieutenants who have just begun their intelligence careers.' "

The latter statement indicates even the masculinity of the intelligence officer is dependent on the espionage process.

Bibliography

[*Maxim's original 3-page Bibliography with its 26 items is omitted here.*]

Indices

Index of Cases—*In Order of Appearance*

This section lists the 60 case studies in the FDDC paper in order of their appearance, each keyed to the chapter, section, and sub-section in which it appears.

CHAPTER	TITLE
1.2.1.	RAF Bomber Command Forgets to Use Deception, Germany 1944.
1.2.2.	British Black Propaganda in the Suez War 1956
1.4.2.1.	Operation HIMMLER: Hitler's Big Lie That Flopped, Poland 1939
1.4.2.2.	German Radio Deception ELEPHANTIASIS Fails to Deter a Russian Attack, 1942
1.4.2.3.	Dakar, British Deception Fails, Sep 1940
1.4.3.1.	Hitler's Pre-invasion of Britain Propaganda Fails, 1940
1.4.3.2.	Allied IRONSIDE Feint at Bay of Biscay, 1944
1.4.3.3.	FORTITUDE NORTH: The Allied Feint against Norway, 1944
1.5.1.1.	Operation MINCEMEAT, Spain 1944
1.5.2.1.	Straw Soldiers, China AD 755
1.5.2.2.	Russian Tactical Radio Deception, Eastern Front 1942
1.5.2.3.	Allied ACCUMULATOR Feint toward the Cotentin Side of Normandy 1944
1.5.3.1.	Jones and the Telltale Decoys, German Occupied France 1944
1.5.3.2.	Bay of Pigs, US Deception Seen Through, Apr 1961
2.1.1.	Col. Clarke, Italian East Africa 1941
2.1.2.	Gen. Alexander at Massicault, Tunisia 1943
2.1.3.	The Errant Airborne, Normandy 1944
2.1.4.	The Score of 23rd Hq Special Troops, Europe 1944-45
2.2.1.	Operation CONDOR, French Indochina 1954

2.2.2.	Operation PURPLE WHALES, Burma 1942
2.2.3.	Col. Fleming's MINCEMEAT-type Plan Fails, Burma 1943
2.2.4.	Dissimulative Camouflage Fails, Germany 1943
2.2.5.	Battle of the Sangro, Italy 1943
2.2.6.	Plan COCKADE, The English Channel 1943
2.2.7.	The Reluctant Pilot, England 1944
2.2.8.	Operation MI, Midway Island 1942
2.2.9.	The German Second Counteroffensive at Anzio, Italy 1944
2.2.10.	Operation ELEPHANT, Normandy 1944
2.2.11.	Twenty Committee Fails at "Coat-Trailing", Germany 1940 43
2.2.12.	Col. Wintle's Device, Cyprus 1942
2.2.13.	MacArthur's Invasion of Luzon, Philippines 1945
2.2.14.	Operation DECOY, Korea 1952
2.3.2.	The German Summer Offensive, Russia 1942
2.3.2.	Operation BREST, France 1944
2.3.3.	The Indaw Raid, Burma 1944
2.3.4.	Dissimulative Camouflage Fails, Japan 1944
2.3.5.	The Tell-Tale Photographs, Germany 1944-45
2.3.6.	The "KGB Poisoners" at Radio Free Europe, Munich 1956
2.3.7.	An R&D Backfire, Czechoslovakia 1960s
2.3.8.	Covert Operations, USA 1987
2.3.9.	Jody Powell and the Iranian Rescue Mission, 1980.
2.3.10.	British Black Propaganda Forgery MÖLDERS Backfires 1941
2.4.1.	Verdun, a German Lure that Turned Around, Western Front 1916
2.4.2.	Dienbienphu, a Disastrous Lure, French Indochina 1953-54
2.5.1.	An "A" Force abort, Cairo 1942
2.5.2.	Aborts by Twenty Committee, England 1941
2.5.3.	The "Italian King's Proclamation" abort, London 1943
2.5.4.	Plan JAEL, London 1943
2.5.5.	Aborts by the 23rd Hq Special Troops, France 1944-45

2.5.6.	Aborts by "D" Division, Burma 1942-45
2.5.7.	Plan RAINBOW, CIA 1956
2.5.8.	R&D Deception Aborts in Czechoslovak Dept. D 1960s
2.6.1.	German Operation ALBION (= SHARK + HARPOON) as Cover for Invasion of Russia 1941
2.6.2.	Operation COPPERHEAD, Gibraltar & Algiers 1944
2.6.3.	Operation ERROR as a Partial Success, Burma 1942
2.7.1.	Operation ERROR as a Potential Failure, Burma 1942
2.7.2.	TRICYCLE-ARTIST-GARBO as a Potential Blown Network, Britain 1944
3.2.1.	Col. Clarke and the Aircraft Struts, Egypt 1942
3.2.2.	"A" Force and the Dummy Tanks, Egypt 1942-43
3.2.3.	Gen. Alexander's Deception Team Double-bluffs, Italy 1944

Index of Cases—*Chronological*

DATE	CHAPTER	PLAN
AD 755	1.5.2.1.	Straw Soldiers, China
1916	2.4.1.	Verdun, a German Lure that Turned Around, Western Front
1939, Aug	1.4.2.1.	Operation HIMMLER: Hitler's Big Lie That Flopped, Poland
1940-43	2.2.11.	Twenty Committee Fails at "Coat-Trailing", Germany
1940, Jul-Sep	1.4.3.1.	Hitler's Pre-invasion of Britain Propaganda Fails
1940, Sep	1.4.2.3.	Dakar, British Deception Fails
1941, Mar	2.6.1.	German Operation ALBION (= SHARK + HARPOON) as Cover for Invasion of Russia
1941, Jan-Dec	2.5.2.	Aborts by Twenty Committee, England
1941, Feb	2.1.1.	Col. Clarke, Italian East Africa
1941, Nov	2.3.10.	British Black Propaganda Forgery MÖLDERS Backfires
1942, early	1.4.2.2.	German Radio Deception ELEPHANTIASIS Fails to Deter a Russian Attack
1942	1.5.2.2.	Russian Tactical Radio Deception, Eastern Front
1942, May	2.6.3.	Operation ERROR as a Partial Success, Burma
1942, May	2.7.1.	Operation ERROR as a Potential Failure, Burma
1942, Jun	2.2.2.	Operation PURPLE WHALES, Burma
1942, Jun	2.2.8.	Operation MI, Midway Island
1942, Jun	2.3.2.	The German Summer Offensive, Russia
1942, mid	3.2.1.	Col. Clarke and the Aircraft Struts, Egypt
1942, Jul	2.2.12.	Col. Wintle's Device, Cyprus
1942-45	2.5.6.	Aborts by "D" Division, Burma
1942, Jul	2.5.1.	An "A" Force abort, Cairo

1942-43	3.2.2.	"A" Force and the Dummy Tanks, Egypt
1943	2.2.4.	Dissimulative Camouflage Fails, Germany
1943, May	2.1.2.	Gen, Alexander at Massicault, Tunisia
1943, Jul	2.5.3.	The "Italian King's Proclamation" abort, London
1943, Sep	2.2.6.	Plan COCKADE, The English Channel
1943, Oct	2.5.4.	Plan JAEL, London
1943, Nov	2.2.3.	Col. Fleming's Haversack-type Plan Fails, Burma
1943, Nov	2.2.5.	Battle of the Sangro, Italy
1944	2.2.7.	The Reluctant Pilot, England
1944, Mar	2.3.3.	The Indaw Raid, Burma
1944, Mar	1.2.1.	RAF Bomber Command Forgets to Use Deception, Germany
1944, Apr	1.5.1.1.	Operation MINCEMEAT, Spain
1944, Feb	2.2.9.	The German Second Counteroffensive at Anzio, Italy
1944, Apr	2.7.2.	TRICYCLE-ARTIST-GARBO as a Potential Blown Network, Britain
1944, May	2.6.1.	Operation COPPERHEAD, Gibraltar & Algiers
1944, May-Jun	1.4.3.1.	Allied IRONSIDE Feint at Bay of Biscay
1944	1.4.3.3.	FORTITUDE NORTH: The Allied Feint against Norway
1944, Jun	2.2.10.	Operation ELEPHANT, Normandy
1944, Jun	2.1.3.	The Errant Airborne, Normandy
1944, Jun	1.5.2.3.	Allied ACCUMULATOR Feint toward the Cotentin Side of Normandy
1944, Aug	1.5.3.1.	Jones and the Telltale Decoys, German Occupied France
1944, Aug	2.3.2.	Operation BREST, France
1944, Aug	3.2.3.	Gen. Alexander's Deception Team Double-bluffs, Italy
1944, Nov	2.3.4.	Dissimulative Camouflage Fails, Japan

1944-45	2.1.4.	The Score of 23rd Hq Special Troops, Europe
1944-45	2.5.5.	Aborts by the 23rd Hq Special Troops, France
1944-45	2.3.5.	The Tell-Tale Photographs, Germany
1945, Jan	2.2.13.	MacArthur's Invasion of Luzon, Philippines
1952	2.2.14.	Operation DECOY, Korea
1953-54	2.4.2.	Dienbienphu, a Disastrous Lure, French Indochina
1954	2.2.1.	Operation CONDOR, French Indochina
1956	2.5.7.	Plan RAINBOW, CIA
1956	2.3.6.	The "KGB Poisoners" at Radio Free Europe, Munich
1956, Oct	1.2.2.	British Black Propaganda in the Suez War
1961, Apr	1.5.3.	Bay of Pigs, US Deception Seen Through
1960s	2.3.7.	An R&D Backfire, Czechoslovakia
1960s	2.5.8.	R&D Deception Aborts in Czechoslovak Dept. D
1980	2.3.9.	Jody Powell and the Iranian Rescue Mission
1987	2.3.8.	Covert Operations, USA

BARTON WHALEY received his bachelor of arts degree in Chinese studies from the University of California, Berkeley, before serving with the intelligence section of U.S. Army Psychological Warfare headquartered in Tokyo during the Korean War. Following the war he attended London University School of Oriental and African Studies before receiving his PhD at Massachusetts Institute of Technology. He was affiliated with the Department of Defense Analysis at the Naval Postgraduate School in Monterey, California, and worked for the director of National Security's Foreign Denial and Deception Committee of the Director of National Intelligence. He passed away in 2013.

The Naval Institute Press is the book-publishing arm of the U.S. Naval Institute, a private, nonprofit, membership society for sea service professionals and others who share an interest in naval and maritime affairs. Established in 1873 at the U.S. Naval Academy in Annapolis, Maryland, where its offices remain today, the Naval Institute has members worldwide.

Members of the Naval Institute support the education programs of the society and receive the influential monthly magazine *Proceedings* or the colorful bimonthly magazine *Naval History* and discounts on fine nautical prints and on ship and aircraft photos. They also have access to the transcripts of the Institute's Oral History Program and get discounted admission to any of the Institute-sponsored seminars offered around the country.

The Naval Institute's book-publishing program, begun in 1898 with basic guides to naval practices, has broadened its scope to include books of more general interest. Now the Naval Institute Press publishes about seventy titles each year, ranging from how-to books on boating and navigation to battle histories, biographies, ship and aircraft guides, and novels. Institute members receive significant discounts on the Press' more than eight hundred books in print.

Full-time students are eligible for special half-price membership rates. Life memberships are also available.

For a free catalog describing Naval Institute Press books currently available, and for further information about joining the U.S. Naval Institute, please write to:

> Member Services
> **U.S. Naval Institute**
> 291 Wood Road
> Annapolis, MD 21402-5034
> Telephone: (800) 233-8764
> Fax: (410) 571-1703
> Web address: www.usni.org